Contractors and Contractor Directions Theory and Applications

LECTURE NOTES
IN PURE AND APPLIED MATHEMATICS

1. *N. Jacobson,* Exceptional Lie Algebras
2. *L.-Å. Lindahl and F. Poulsen,* Thin Sets in Harmonic Analysis
3. *I. Satake,* Classification Theory of Semi-Simple Algebraic Groups
4. *F. Hirzebruch, W. D. Newmann, and S. S. Koh,* Differentiable Manifolds and Quadratic Forms
5. *I. Chavel,* Riemannian Symmetric Spaces of Rank One
6. *R. B. Burckel,* Characterization of C(X) Among Its Subalgebras
7. *B. R. McDonald, A. R. Magid, and K. C. Smith,* Ring Theory: Proceedings of the Oklahoma Conference
8. *Y.-T. Siu,* Techniques of Extension of Analytic Objects
9. *S. R. Caradus, W. E. Pfaffenberger, and B. Yood,* Calkin Algebras and Algebras of Operators on Banach Spaces
10. *E. O. Roxin, P.-T. Liu, and R. L. Sternberg,* Differential Games and Control Theory
11. *M. Orzech and C. Small,* The Brauer Group of Commutative Rings
12. *S. Thomeier,* Topology and Its Applications
13. *J. M. López and K. A. Ross,* Sidon Sets
14. *W. W. Comfort and S. Negrepontis,* Continuous Pseudometrics
15. *K. McKennon and J. M. Robertson,* Locally Convex Spaces
16. *M. Carmeli and S. Malin,* Representations of the Rotation and Lorentz Groups: An Introduction
17. *G. B. Seligman,* Rational Methods in Lie Algebras
18. *D. G. de Figueiredo,* Functional Analysis: Proceedings of the Brazilian Mathematical Society Symposium
19. *L. Cesari, R. Kannan, and J. D. Schuur,* Nonlinear Functional Analysis of Differential Equations: Proceedings of the Michigan State University Conference
20. *J. J. Schäffer,* Geometry of Spheres in Normed Spaces
21. *K. Yano and M. Kon,* Anti-Invariant Submanifolds
22. *W. V. Vasconcelos,* The Rings of Dimension Two
23. *R. E. Chandler,* Hausdorff Compactifications
24. *S. P. Franklin and B. V. S. Thomas,* Topology: Proceedings of the Memphis State University Conference
25. *S. K. Jain,* Ring Theory: Proceedings of the Ohio University Conference
26. *B. R. McDonald and R. A. Morris,* Ring Theory II: Proceedings of the Second Oklahoma Conference
27. *R. B. Mura and A. Rhemtulla,* Orderable Groups
28. *J. R. Graef,* Stability of Dynamical Systems: Theory and Applications
29. *H.-C. Wang,* Homogeneous Banach Algebras
30. *E. O. Roxin, P.-T. Liu, and R. L. Sternberg,* Differential Games and Control Theory II
31. *R. D. Porter,* Introduction to Fibre Bundles
32. *M. Altman,* Contractors and Contractor Directions Theory and Applications

Other Volumes in Preparation

Contractors and Contractor Directions Theory and Applications

A NEW APPROACH TO SOLVING EQUATIONS

Mieczyslaw Altman
Department of Mathematics
Louisiana State University
Baton Rouge, Louisiana

MARCEL DEKKER, INC. New York and Basel

Library of Congress Cataloging in Publication Data

Altman, Mieczyslaw, 1916-
Contractors and contractor directions theory and applications.

(Lecture notes in pure and applied mathematics ; 32)
Includes bibliographical references and index.
1. Numerical analysis. 2. Equations--Numerical solutions. 3. Functional analysis. I. Title.
II. Title: Contractor directions theory and applications.
QA297.A54 519.4 77-21442
ISBN 0-8247-6672-5

MARCEL DEKKER, INC.

270 Madison Avenue, New York, New York 10016

Current printing (last digit):
10 9 8 7 6 5 4 3 2 1

PRINTED IN THE UNITED STATES OF AMERICA

To

Wanda*Tom*Barbara

PREFACE

This monograph is an outgrowth of research which was started in 1971. The main topic, which is as old as the history of recorded mathematics, is solving equations by analytical means. Our aim in the monograph is to attempt to present a general theory which will enable us to unify the largest possible variety of basic theoretical results dealing with solving nonlinear equations. For this purpose the concept of contractors and contractor directions provides a powerful tool for building a bridge over methods and theories which initially seemed to be far apart. Upon examination, using this concept, a great number of well-known procedures and algorithms, including the most important ones, which had appeared to be so very different in nature and character, turned out to be particular cases of a unified approach.

As a result of our investigations this monograph reports our latest conclusions, published now for the first time. New material will be found here, for example, on contractors and secant methods for solving equations (Chapter 7), contractors and roots of nonlinear functions (Chapter 8), relations research (Chapter 10), and contractors, approximate identities and factorization in Banach algebras (Appendix 3, Sections 6 and 7). While the book contains no background material, the reader familiar with a first course in Functional Analysis will be able to grasp the main thrust of the text.

I gave a course on the theory of contractors at the University of Newcastle, New South Wales, Australia during the winter semester in 1973. Shortly after, I repeated the course at the Louisiana State University, Baton Rouge.

At this point, my thanks are due to the University of Newcastle for its hospitality and research facilities, to Louisiana State University for the opportunity to continue my research, and to Monica Loftin for her excellent typing of the monograph and numerous research papers which have contributed to the manuscript.

I owe special thanks to Marcel Dekker, Inc., for making available this new development in mathematical research by publishing the monograph in this series.

Miecyzslaw Altman

CONTENTS

INTRODUCTION

Considerable efforts to use modern concepts of Functional Analysis have been made by various mathematicians in order to build up general theories of solving equations. To mention a few: Kantorovich [1] has constructed a general theory for approximate solutions of linear operator equations. A systematic extensive study of iterative procedures is contained in the monograph by Ortega and Rheinboldt [2] (see also Rheinboldt [1]).

In the monograph by Krasnosel'skii and his school [2], special emphasis is placed on nonlinear problems. Recently, Petryshyn [1] developed a very general theory of A-proper operators for constructive approximate solutions of linear and nonlinear equations.

This monograph is a recapitulation of our published, submitted and unpublished results obtained since 1971. Two new concepts are basic for the entire theory: the concept of contractors and the concept of contractor directions. The notions of contractors and nonlinear majorant functions can be easily combined to give both existence and convergence theorems. These results which also reveal the character of the convergence and provide error estimates, are the basis for a unified theory for a large class of iterative methods, including the most important ones: the method of successive approximations, the Newton-Kantorovich method, Newton's method for nonlinear functionals, the method of steepest descent and other gradient type methods. All these various methods are evidently different in nature but can be unified in a general theory by a single concept of contractors which is rather surprising. This fact does not mean to "generalize for the sake of generalization." It is often the case that new concepts create new techniques and applications. For instance, recently Yoshimura [1] gave an application of the contractor method to the Quantum Field Theory. A further development of the contractor technique is aimed at including various two point secant methods, in particular, methods of Seffensen's type.

On the other hand, the method of contractor directions is semi-iterative in the sense that we have iterations with small steps. To handle such iterations, we have considerably developed a transfinite induction technique based upon the idea which was used by Gavurin [1] in a particular case. The method of contractor directions yields ultimately to a new very general solvability principle for nonlinear abstract equations. In this way, some recent results can be obtained as particular cases of our general theory. The method of contractor directions can also be applied to nonlinear differential and integral equations. Moreover, the notion of a contractor is introduced in Banach algebras in close connection with approximate identities and factorization problems.

CHAPTER 1

INVERSE DIFFERENTIABILITY AND CONTRACTORS

Introduction. Recently, Nashed [1] gave a systematic and very comprehensive exposition of abstract differential calculus in normed and topological linear spaces showing the important role of differentials in nonlinear functional analysis. Yamamuro [1] discussed various definitions of differentiability with main focus on the weakest and strongest derivatives. In this chapter, we are concerned with one aspect of this concept as a tool to investigate iteration procedures for solving equations in Banach spaces. It is the nature of many applied problems leading to operator equations that the inverse operator is required to exist and to be even continuous. Thus, the regularity conditions are in the range space. For this reason for instance, the iteration procedure in the well known implicit function theorem uses the inverse of the Fréchet (Gâteaux) derivative. The same fact is seen in the Newton-Kantorovič procedure (see Kantorovič and Akilov [1], Collatz [1], Ortega and Rheinboldt [2]).

This observation leads us to an independent definition of an inverse derivative such that can be used in place of the inverse of the derivative. Moreover, it turns out that for the same purposes much less can be required and what is the concept of a contractor. As a matter of fact if the mapping is a contraction itself, then the contractor exists and is simply the identity mapping. The concept of a contractor generalizes the notion of an inverse derivative.

1. Inverse derivatives

Let $P:X \to Y$ be a nonlinear operator from a Banach space X to a Banach space Y. Consider the difference $P(x+h)-Px = Q(x)h$ and suppose that $\Gamma(x)$ is a linear bounded operator associated with $x \in X$ acting from Y to X, i.e., $\Gamma(x):Y \to X$.

If $\Gamma(x)$ has the property

$$(1.1) \qquad \|y\|^{-1}\|Q(x)\Gamma(x)y - y\| \to 0 \quad \text{as} \quad y \to 0, \quad \text{for every} \quad y \in Y,$$

then $\Gamma(x)$ is called the inverse derivative at x of P. Condition (1.1) can be written in the form

$$(1.2) \qquad \|y\|^{-1}\|P(x+\Gamma(x)y) - Px - y\| \to 0 \quad \text{as} \quad y \to 0 .$$

Properties of inverse derivatives

(i) If $\Gamma(x)$ exists, then $\frac{d}{dt} P(x+t\Gamma(x)y)\Big|_{t=0} = y$, i.e., P has a directional derivative in the direction $\Gamma(x)y$.

Property (i) shows that

(ii) $\Gamma(x)y = 0$ implies $y = 0$, i.e., $\Gamma(x)$ is a one-to-one mapping.

(iii) If the Fréchet derivative $P'(x)$ exist, then $P'(x)$ is an extension of $[\Gamma(x)]^{-1}$.

(iv) If $[P'(x)]^{-1}$ exists, then $\Gamma(x)$ is uniquely defined, and $\Gamma(x) = [P'(x)]^{-1}$.

(v) If $\Gamma(s)$ is onto, then $P'(x)$ exists and has the inverse $[P'(x)]^{-1} = \Gamma(x)$.

Proposition. If $P'(x)$ and $\Gamma(x)$ exist, then $P'(x)$ maps X onto Y, and there is a projection of X onto the kernel of $P'(x)$, i.e., $X = N \oplus X_N$ (direct sum), where $N = [h : P'(x)h = 0, h \in X]$, and X_N is the range of $\Gamma(x)$.

Proof. Since $P'(x)\Gamma(x)y = y$, for arbitrary $y \in Y$, by (iii), $P'(x)$ is onto. It is easily seen that X_N is closed. In fact, $\Gamma(x)y_n \to x_0$ implies $y_n = P'(x)\Gamma(x)y_n \to P'(x)x_0 = y_0$, by (iii). Hence, $\Gamma(x)y_n \to \Gamma(x)y_0 = x_0$.

For an arbitrary h of X, let $y = P'(x)h$ and $h_p = \Gamma(x)y$.

Then, by (iii), $P'(x)h_p = [\Gamma(x)]^{-1}h_p = y$. Hence, $P'(x)(h-h_p) = 0$, i.e., $h-h_p \in N$, where $h_p \in X_N = \Gamma(x)(Y)$. Clearly, $N \cap X_N = \{0\}$.

Note that this concept of an inverse derivative is considered in the strong (Frêchet) sense. However, it is also possible to introduce the notion of an inverse derivative in a weaker sense of Gâteaux.

Remark 1.1. Calling $\Gamma(x)$ a right derivative, we could also introduce, in a similar way, a left inverse derivative $\Gamma(x)$ using the following formula in place of (1.1),

$$\|h\|^{-1}\|\Gamma(x)Q(x)h-h\| \to 0 \quad \text{as} \quad h \to 0 , \quad h \in X .$$

2. Iteration procedures with inverse derivatives

Our problem is to find a solution to the operator equation

$$Px = 0 , \tag{2.1}$$

where $P:X \to Y$, X and Y are Banach spaces. We assume the existence of inverse derivatives $\Gamma(x)$ of P in a neighborhood $S(x_0,r) = [x : \|x-x_0\| \leq r , x \in X]$, where x_0 is a given approximate solution to (2.1). For solving (2.1), we use the following iteration procedure,

$$x_{n+1} = x_n - \Gamma(x_n)Px_n , \quad n = 0,1,2,\cdots . \tag{2.2}$$

The following theorem gives sufficient conditions for the convergence of the iteration procedure (2.2) to a solution of equation (2.1).

Theorem 2.1. Suppose that there exist positive numbers $0 < q < 1$, r , η and B such that the inverse derivative satisfies the uniformity condition.

$$\|y\|^{-1}\|P(x+\Gamma(x)y) - Px - y\| \leq q \quad \text{for} \quad x \in S(x_0,r) \quad \text{and} \quad \|y\| \leq \eta . \tag{2.3}$$

$$\|\Gamma(x)\| \leq B \quad \text{for} \quad x \in S(x_0,r) . \tag{2.4}$$

$$\|P(x_0)\| \leq \eta . \tag{2.5}$$

(2.6) $$B\eta(1-q)^{-1} \leq r .$$

(2.7) P is closed on $S(x_0,r)$.†

Then there exists a solution $x^* \in S(x_0,r)$ and the sequence of x_n defined by (2.2) converges toward x^* , i.e.,

$$x_n \to x^* , \quad Px^* = 0 , \quad x^* \in S(x_0,r)$$

and

(2.8) $$\|x_n - x^*\| \leq B\,\eta q^n (1-q)^{-1} .$$

Proof. Putting $y = -Px_n$ in (2.3), we obtain

(2.9) $$\|Px_{n+1}\| \leq q\|Px_n\| , \quad n = 0,1,2,\cdots .$$

Hence, it follows from (2.2) and (2.3) - (2.5),

(2.10) $$\|x_{n+1} - x_n\| \leq B\|Px_n\| \leq B\,\eta q^n .$$

By induction, it is easy to see that $x_n \in S(x_0,r)$ for $n = 0,1,2,\cdots$. Thus the sequence of x_n converges to some $x^* \in S(x_0,r)$. Since $Px_n \to 0$, by (2.9), and P is closed on $S(x_0,r)$, it follows that $Px^* = 0$. The error estimate (2.8) results from (2.10) in the usual way.

Remark 2.1. If $\Gamma(x)$ is onto, then the iteration procedure (2.2) becomes the well known Newton-Kantorovič method (see Kantorovič and Akilov [1], Collatz [1], Ortega and Rheinboldt [2]).

(2.11) $$x_{n+1} = x_n - [P'(x_n)]^{-1} Px_n .$$

In this case, under the hypotheses of Theorem 2.1, the solution x^* is unique in $S(x_0,r)$ if, in addition, condition (2.3) is satisfied for all $y \in Y$ such that $\|\Gamma(x)y\| \leq 2r$, where $x \in S(x_0,r)$.

† i.e., $x_n \in S$, $x_n \to x$ and $Px_n \to y$ imply that $x \in S$ and $y = Px$.

This results from the following inequality,

$$(1-q)\|y\| < \|P(x+\Gamma(x)y) - Px\| \tag{2.12}$$

obtained from (2.3). For if x^*, $x^{**} \in S(x_0,r)$ are two solutions, then x^* can be written as $x^{**} = x^*+\Gamma(x^*)y$ and we have

$$\|\Gamma(x^*)y\| = \|x^{**}-x^*\| \leq 2r$$

and we can apply the inequality (2.12).

Consider now the equation

$$F(x) = 0 ,$$

where $F:X \to R$ (reals) is a nonlinear functional on X. If the gradient $F'(x)$ exists and $F'(x)h \neq 0$, where $h \in X$, then we have

$$t^{-1}|F(x+t[F'(x)h]^{-1}h) - F(x) - t| \to 0 \quad \text{as} \quad t \to 0 ,$$

where $t \in R$, i.e., the mapping from R to X defined by

$$\Gamma(x)t = t[F'(x)h]^{-1}h$$

is an inverse derivative of F at x, if h is fixed and $F'(x)h \neq 0$, by virtue of (1.2).

Consider now the following generalization of Newton's method for nonlinear functionals which is given by Altman [18],

$$x_{n+1} = x_n - [F'(x_n)h_n]^{-1}F(x_n)h_n \; ; \quad x_n , \; h_n \in X . \tag{2.13}$$

<u>Remark 2.2</u>. The generalized Newton method (2.13) for nonlinear functionals is also a special case of an iteration procedure (2.2) with inverse derivatives.

Although the methods (2.11) and (2.13) are entirely different, both can be considered as particular cases of the procedure (2.2).

We say that $\Gamma(x_0)$ is a uniform inverse derivative of P at x_0 if the following condition is satisfied (provided $\Gamma(x_0)$ is an inverse derivative),

$$\|y\|^{-1}\|P(x+\Gamma(x_0)y) - Px - y\| \le q \quad \text{for} \quad \|y\| \le \eta$$

and for x in some neighborhood of x_0. Using this notion, we consider the following modification of procedure (2.2),

(2.14) $$x_{n+1} = x_n - \Gamma(x_0)Px_n , \quad n = 0,1,2,\cdots .$$

Theorem 2.2. Suppose that there exist positive numbers $0 < q < 1$, r, η and B such that the uniform inverse derivative $\Gamma(x_0)$ satisfies the condition

(2.15) $$\|y\|^{-1}\|P(x+\Gamma(x_0)y) - Px - y\| \le q \quad \text{for} \quad x \in S(x_0,r)$$

and

$$\|y\| \le \eta \|\Gamma(x_0)\| \le B \quad \text{and} \quad \|Px_0\| \le \eta .$$

Then there exists a solution $x^* \in S(x_0,r)$ and the sequence of x_n determined by (2.14) converges toward x^*, i.e., $x_n \to x^*$, $Px^* = 0$, $x^* \in S(x_0,r)$ and the error estimate (2.8) holds. If $\Gamma(x_0)$ is onto and (2.15) is satisfied for all $y \in Y$ such that $\|\Gamma(x_0)y\| \le 2r$, then the solution x^* is unique in $S(x_0,r)$.

The proof of this theorem is exactly the same as that of Theorem 2.1.

3. Contractors

Analyzing the proof of Theorem 2.1, we can see that condition (2.3) plays the basic role in our argument. This observation leads to the concept of a contractor. Let $P:X \to Y$ be a nonlinear mapping and let $\Gamma(x):Y \to X$ be a bounded linear operator associated with x.

Definition 3.1. We say P has a contractor $\Gamma(x)$ if there is a positive number $q < 1$ such that

(3.1) $$\|P(x+\Gamma(x)y) - Px - y\| \le q\|y\| ,$$

where $x \in X$ and $y \in Y$ are defined by the particular problem.

Example 3.1. If P has a Fréchet derivative $P'(x)$ satisfying

$$(3.1^*) \qquad \|P'(x)\Gamma(x)y - y\| \leq q'\|y\| \ , \quad 0 < q' < 1 \ ,$$

where $\Gamma(x): Y \to X$ is a linear bounded operator, then $\Gamma(x)$ is a contractor and condition (3.1) is satisfied with $q = (1+q')/2$ and $\|y\| \leq \delta\|\Gamma(x)\|^{-1}$, where δ is chosen so as to satisfy

$$\|P(x+h) - Px - P'(x)h\| \leq (1-q')2^{-1}\|h\| \cdot \|\Gamma(x)\|^{-1}$$

for $\|h\| \leq \delta$. Then for $h = \Gamma(x)y$, we have

$$\|P(x+\Gamma(x)y - Px - y\| \leq \|P(x+\Gamma(x)y) - Px - P'(x)\Gamma(x)y\| +$$

$$+ \|P'(x)\Gamma(x)y - y\| \leq [(1-q')2^{-1} + q']\|y\| \ , \quad \text{and} \quad \|y\| \leq \delta\|\Gamma(x)\|^{-1}$$

$$\text{implies} \quad \|h\| = \|\Gamma(x)y\| \leq \delta \ .$$

Example 3.2. An inverse derivative is a contractor.

We say that $P: X \to Y$ has a bounded contractor $\Gamma(x)$ if $\|\Gamma(x)\| \leq B$ for all x of a certain region.

Example 3.3. Assume that $P: X \to Y$ has a Fréchet derivative $P'(x)$ which is Hölder continuous, i.e., there exist positive numbers K, $\alpha \leq 1$ such that

$$\|P'(x) - P'(\bar{x})\| \leq K\|x - \bar{x}\|^{\alpha} \quad \text{for all} \quad x, \bar{x} \in X \ .$$

Suppose that $\Gamma(x)$ is bounded by B, i.e.,

$$\|\Gamma(x)\| \leq B \quad \text{for all} \quad x \in X \ .$$

and satisfies condition (3.1^*). Then $\Gamma(x)$ is a bounded contractor for P and satisfies inequality (3.1) with arbitrary q such that $q' < q < 1$ and for all $y \in Y$ such that

$$\|y\| \leq r = [(1+\alpha)(q-q')]^{1/\alpha}K^{-1}B^{-(1+1/\alpha)} \ .$$

Proof. We have

$$\|P(x+\Gamma(x)y) - Px - y\| \leq \|P(x+\Gamma(x)y) - Px - P'(x)\Gamma(x)y\| +$$

$$+ \|P'(x)\Gamma(x)y - y\| \leq K(1+\alpha)^{-1}\|\Gamma(x)y\|^{1+\alpha} + q'\|y\| \leq$$

$$\leq [K(1+\alpha)^{-1}B^{1+\alpha}\|y\|^{\alpha} + q']\|y\| \leq q\|y\| \quad \text{if} \quad \|y\| \leq r .$$

Example 3.4. Let $F:X \to X$ be a contraction mapping, i.e., there exists a positive number $q < 1$ such that

$$\|Fx-F\bar{x}\| \leq q\|x-\bar{x}\| \quad \text{for all} \quad x,\bar{x} \in X .$$

Then the operator P of the form $Px = x-Fx$, $x \in X$ has $\Gamma(x) \equiv I$ (identity mapping) as a contractor. In fact, we have

$$\|P(x+\Gamma(x)y) - Px - y\| = \|x+\Gamma(x)y - F(x+\Gamma(x)y) - x + Fx - y\| =$$

$$= \|Fx - F(x+y)\| \leq q\|y\| \quad \text{for all} \quad y \in X .$$

Suppose that $P:X \to Y$ has a contractor $\Gamma(x)$ satisfying (3.1) for all $y \in Y$. Then it is easily seen that the following inequality can be derived from (3.1),

$$(1-q)\|y\| \leq \|P(x+\Gamma(x)y) - Px\| \quad \text{for} \quad y \in Y . \tag{3.2}$$

It follows from (3.1) that the contractor $\Gamma(x)$ is a one-to-one mapping and if P is continuous, then (3.2) yields the continuity of the inverse $[\Gamma(x)]^{-1}$, i.e., then $\Gamma(x)$ is a homeomorphism of Y onto a closed subspace of X. A contractor $\Gamma(x)$ is called regular, if (3.1) is satisfied for all $y \in Y$ and $D(P) = \Gamma(x)(Y)$, where $D(P)$ is the domain of P. We say that x is a regular point of P, if $P^{-1}(Px) = \{x\}$ and $Px_n \to Px$ implies $x_n \to x$.

Lemma 3.1. If a contractor $\Gamma(x)$ exists for $x \in D(P)$ and is regular, then x is a regular point of P. If $\Gamma(x)$ is regular and onto, i.e., $\Gamma(x)(Y) = X$, then X is a regular point of P and P

is continuous at x . If $\Gamma(x)$ is regular for every $x \in D(P)$, then P has a continuous inverse mapping P^{-1} . If $\Gamma(x)$ is regular and onto for every $x \in D(P)$, then P is a homeomorphism of X onto $P(X)$.

Proof. The proof follows from inequality (3.2), since $x_n \in D(P)$ implies $x_n - x = \Gamma(x)y_n$ for some $y_n \in Y$. If $\Gamma(x)$ is onto, then the continuity of P results from the continuity of the inverse operator $[\Gamma(x)]^{-1}$ and from (3.1).

Theorem 3.1. Theorem 2.1 remains true if we replace there the inverse derivative by a bounded contractor $\Gamma(x)$.

The proof is the same as that of Theorem 2.1.

We say that the linear bounded operator $\Gamma(x_0):Y \to X$ is a uniform contractor of P at x_0 if there exist positive numbers $0 < q < 1$, r and η such that condition (2.15) is satisfied for $x \in S(x_0,r)$ and $\|y\| \leq \eta$. If the Fréchet derivative $P'(x)$ exists and is Lipschitz continuous with constant K in some neighborhood of x_0 and $P'(x_0)$ is invertible, then $\Gamma(x_0) = [P'(x_0)]^{-1}$ is a uniform inverse derivative, since

$$\|P(x+\Gamma(x_0)y) - Px - y\| = \|P(x+\Gamma(x_0)y) - Px - P'(x_0)\Gamma(x_0)y\| \leq$$

$$\leq \|P(x+\Gamma(x_0)y) - Px - P'(x)\Gamma(x_0)y\| + \|P'(x)\Gamma(x_0)y - P'(x_0)\Gamma(x_0)y\| \leq$$

$$\leq 2^{-1}K\|\Gamma(x_0)\|^2\|y\|^2 + K\|x-x_0\|\|\Gamma(x_0)\|\|y\| .$$

Evidently, $\Gamma(x_0)$ is a uniform contractor.

Theorem 3.2. Theorem 2.2 remains true if we replace there the uniform inverse derivative by the uniform contractor $\Gamma(x_0)$.

The proof of this theorem is exactly the same as that of Theorem 2.2.

4. Implicit function theorems using contractors

On the basis of Theorems 3.1 and 3.2 we can generalize the well known implicit function theorem. Let X, Z and Y be Banach spaces. Consider the operator $P(x,z): X \times Z \to Y$. Put $S = [(x,z) : \|x-x_0\| \leq r, \|z-z_0\| \leq \rho]$ and suppose that, for every z such that $(x,z) \in S$, P has a contractor $\Gamma(x,z): Y \to X$ which is strongly continuous in z for each fixed x.

Theorem 4.1. Suppose that there exist positive numbers $0 < q < 1$, r, ρ, η and B such that

(4.1) $\|P(x+\Gamma(x,z)y,z) - P(x,z) - y\| \leq q\|y\|$ for $(x,z) \in S$ and $\|y\| \leq \eta$.

(4.2) $$\|P(x,z)\| \leq B \quad \text{for} \quad (x,z) \in S .$$

(4.3) $$\|P(x_0,z)\| \leq \eta .$$

(4.4) $$B\eta(1-q)^{-1} \leq r .$$

(4.5) $P(x,z)$ is closed on S for every fixed z restricted to S, i.e., $(x_n,z) \in S$, $x_n \to x$ and $P(x_n,z) \to y$ imply $y = P(x,z)$.

(4.6) $P(x,z)$ is continuous in S with respect to z for every fixed x restricted to S.

(4.7) $$P(x_0,z_0) = 0 .$$

Then there exists a continuous function $x = g(z)$ such that $P(g(z),z) = 0$, $(g(z),z) \in S$.

Proof. The proof is based on Theorem 3.1. Consider the iteration procedure $x_{n+1}(z) = x_n(z) - \Gamma(x_n,z)P(x_n,z)$ for $n = 0,1,2,\cdots$ and every fixed z restricted to S. By virtue of Theorem 3.1, the sequence of $x_n(z)$ converges to an element $x = g(z)$ such that $(x,z) \in S$ and $P(x,z) = 0$. The functions $x_n(z)$ are continuous and it is easily seen that the convergence of $x_n(z)$ is uniform in z. Thus, $x(z)$ is continuous.

Remark 4.1. If $P(x,z)$ is jointly continuous in (x,z), then conditions (4.1), (4.2) and (4.7) are sufficient for the theorem and numbers r, ρ can be chosen so as to satisfy (4.3) and (4.7). If $\Gamma(x,z)$ is onto, then the function $g(z)$ is unique in S, provided condition (4.1) is satisfied for all $y \in Y$ such that $\|\Gamma(x,z)y\| \leq 2r$.

Theorem 4.2. Theorem 4.1 remains true if we replace there the contractor $\Gamma(x,z)$ by a uniform contractor $\Gamma(x_0,z)$, i.e., if we replace condition (4.1) by

$$\|P(x+\Gamma(x_0,z)y,z) - P(x,z) - y\| \leq q\|y\| .$$

Remark 4.1 remains also true.

The proof is exactly the same as that of Theorem 4.1, but we use here the following iteration procedure

$$x_{n+1}(z) = x_n(z) - \Gamma(x_0,z)P(x_n,z) \quad \text{for} \quad n = 0,1,2,\cdots .$$

5. A generalization of the Banach contraction principle

Consider the operator equation $Px = \xi$, where $P:D(P) \subset X \to Y$ is a closed nonlinear operator.

Theorem 5.1. Suppose that the closed operator P has a bounded contractor Γ such that

$$\|P(x+\Gamma(x)y) - Px - y\| \leq q\|y\| \quad \text{for all } y \in Y , \tag{5.1}$$

where $0 < q < 1$ and $\|\Gamma(x)\| \leq B$ for all $x \in D(P)$. Then the equation $Px = y$ has a solution for arbitrary $y \in Y$. If $\Gamma(x)$ is regular (and onto) for every $x \in D(P)$, then the inverse P^{-1} exists and is Lipschitz continuous with the constant $B(1-q)^{-1}$ (and P is continuous).

Proof. For arbitrary fixed $\xi \in Y$ consider the operator determined by $Px - \xi$. This operator has the same contractor $\Gamma(x)$ and we apply the iteration procedure

(5.2) $$x_{n+1} = x_n - \Gamma(x_n)(Px_n - \xi) ,$$

where $x_0 \in D(P)$ is arbitrarily chosen. The sequence of x_n converges to an element $x = R\xi$ and $Px_n \to \xi$ as $n \to \infty$. Since P is closed, $P(R\xi) = \xi$.

Suppose now that $\Gamma(x)$ is regular. Let $P(R\xi_1) = \xi_1$ and $P(R\xi_2) = \xi_2$. Then we can write $R\xi_2 = R\xi_1 + \Gamma(R\xi_1)y$. Hence, using inequality (3.2), we obtain

$$\|R\xi_2 - R\xi_1\| \le \|\Gamma(R\xi_1)\|\|y\| \le B(1-q)^{-1}\|\xi_2 - \xi_1\| .$$

From Lemma 3.1 follows that if $\Gamma(x)$ is onto, then P is continuous.

The global constant B can be replaced by a local one $B(x_0,\xi)$ such that $\|\Gamma(x)\| \le B(x_0,\xi)$ for $\|x-x_0\| \le r$ and $B(x_0,\xi)\|Px_0 - \xi\|(1-q)^{-1} \le r$. Then we can prove only the continuity of P^{-1}.

Consider now equations of the second kind $Px = x - Fx = \xi$, where $F: X \to X$. Condition (5.1) yields here

(5.3) $$\|Fx - F(x+\Gamma(x)y) - (I-\Gamma(x))y\| \le q\|y\| \quad \text{for all} \quad y \in X ,$$

where I is the identity mapping.

<u>Theorem 5.2</u>. Theorem 5.1 remains true for the equation $x - Fx = \xi$, $x,\xi \in X$, if we replace there condition (5.1) by (5.3).

The proof follows from Theorem 5.1.

<u>Remark 5.1</u>. If $F: X \to X$ is a contraction with the constant $q < 1$, then, obviously, a contractor $\Gamma(x)$, satisfying all conditions of Theorem 5.2 exists and it is the identity mapping, i.e., $\Gamma(x) = I$. Thus, Theorem 5.2 generalizes the well known Banach contraction principle.

The following remark will be used in the next section.

Remark 5.2. Consider the equation

$$Px = x - F(x) = \xi \;, \quad x, \xi \in X \;. \tag{5.4}$$

Then Theorem 5.2 remains true if condition (5.3) is replaced by the following inequality

$$\|F(x+y+\tilde{\Gamma}(x)y) - Fx - \tilde{\Gamma}(x)y\| \le q\|y\| \;, \quad \text{for all} \quad y \in X \tag{5.5}$$

and $0 < q < 1$, where $\|\tilde{\Gamma}(x)\| \le B$.

For we replace in (5.2) the operator $I - \Gamma(x)$ by $\tilde{\Gamma}(x)$. If $(I + \tilde{\Gamma}(x))(X) = D(F)$ for all $x \in D(F)$ then the inverse of $x - F(x)$ exists and is defined on the whole of X and is Lipschitz continuous with the constant $(1+B)(1-q)^{-1}$. The iteration procedure in this case is determined as follows

$$x_{n+1} = x_n - (I + \tilde{\Gamma}(x_n))(x_n - Fx_n - \xi) \;, \quad n = 0,1,2,\cdots \;. \tag{5.6}$$

The initial approximate solution x_0 can be chosen arbitrarily and the procedure converges toward a solution. For equations of second kind, it is convenient to have the contractor in the form $I + \tilde{\Gamma}(x)$, $x \in D(F)$.

6. Nonlinear evolution equations, a generalization of the Piccard theorem in Banach spaces

Consider the initial value problem

$$\frac{dx}{dt} = F(t,x) \;, \quad 0 \le t \le T \;, \quad x(0) = \xi \;, \tag{6.1}$$

where the unknown $x = x(t)$ is a function defined on the real interval $[0,T]$ with values in the Banach space X, and $F: [0,T] \times X \to X$ is a continuous mapping. Instead of (6.1) we consider the integral equation

$$x(t) - \int_0^t F(s,x(s))ds = \xi \;. \tag{6.2}$$

Denote by X_T the space of all continuous functions $x = x(t)$

defined on $[0,T]$ with values in X and with the norm $\|x\|_C = \max[\|x(t)\| : 0 \leq t \leq T]$. Considering equation (6.2) as an operator equation in X_T , we can apply our generalization of the Banach contraction principle discussed in Section 5, especially Remark 5.2.

For arbitrary fixed $x \in X$ and $t \in [0,T]$, let $\Gamma(t,x):X \to X$ be a bounded linear operator, strongly continuous with respect to (t,x) in the sense of the operator norm. Suppose that there exist positive numbers K , B such that the inequality

$$
\begin{aligned}
&\max_{0 \leq t \leq T} \|F(t,x(t)+y(t) + \int_0^t \Gamma(s,x(s))y(s)ds) - \\
(6.3) \qquad &- F(t,x(t)) - \Gamma(t,x(t))y(t)\| \leq K\|y\|_C
\end{aligned}
$$

is satisfied for arbitrary continuous functions $x = x(t)$, y
$y = y(t) \in X_T$, where

$$\|\Gamma(t,x)\| \leq B \quad \text{for all} \quad x \in X \quad \text{and} \quad t \in [0,T] \ .$$

Then we say that $F(t,x)$ has a bounded integral contractor $\{I + \int_0^t \Gamma\}$. A bounded integral contractor is said to be regular if the integral equation

$$(6.4) \qquad y(t) + \int_0^t \Gamma(s,x(s))y(s)ds = z(t) \ , \quad 0 \leq t \leq T$$

has a continuous solution $y(t)$ for arbitrary fixed and continuous functions $x(t)$ and $z(t) \in X_T$. Obviously, if $F(t,x)$ satisfies Lipschitz condition uniformly in t , then $\{I + \int_0^t \Gamma\}$, where $\Gamma \equiv 0$, is a regular bounded integral contractor.

Theorem 6.1. Suppose that $F(t,x)$ has a bounded integral contractor and T is such that $TK = q < 1$. Then, for arbitrary $\xi \in X$ equation (6.2) has a continuous solution $x(t)$. If the bounded integral contractor is regular, then the solution $x \in X_T$ is unique and Lipschitz continuous with respect to ξ .

Proof. Consider in X_T the iteration procedure

$$x_{n+1} = x_n - [y_n + \int_0^t \Gamma(s,x_n)y_n ds] \quad \text{for} \quad n = 0,1,2,\cdots , \tag{6.5}$$

where

$$x_n = x_n(t) \ , \quad y_n = y_n(t) = x_n(t) - \int_0^t F(s,x_n(s))ds - \xi \ .$$

We have, by (6.5),

$$y_{n+1} = \int_0^t [F(s,x_n) - F(s,x_n - y_n - \int_0^s \Gamma(\tau,x_n)y_n d\tau) - \Gamma(s,x_n)y_n] ds \ .$$

Hence, replacing y by $-y$ in (6.3), we obtain that $\|y_{n+1}\|_C \leq$ $\leq q\|y_n\|_C$. Thus, the sequence of y_n is convergent in X_T toward zero. Since $\|\Gamma(t,x)\| \leq B$, it follows from (6.5) that $\sum_{n=0}^{\infty} \|x_{n+1}-x_n\|_C \leq (1+TB)\|y_0\|_C(1-q)^{-1}$, and that $x_n = x_n(t)$ is continuous for $n = 0,1,2,\cdots$, where $x_0 = x_0(t) \in X_T$ can be arbitrarily chosen. Therefore, the sequence of x_n has a limit $x \in X_T$ which is a solution to (6.2). In the same way as in the proof of Theorem 5.2, we derive from (6.4) that the solution x is unique and Lipschitz continuous with respect to ξ .

It is interesting to observe that the contractor for (6.2) at each fixed (t,x) is naturally defined as a linear integral operator of the same kind as the operator in the equation (6.2).

Theorem 6.1 generalizes the well known Piccard theorem for evolution equations in Banach spaces (see Kato [1]). For as we mentioned above, if $F(t,x)$ is t-uniformly Lipschitz continuous, then with $\Gamma \equiv 0$ it satisfies the assumptions of Theorem 6.1.

7. A coincidence theorem and a generalization of Krasnosel'skii's fixed point theorem

The purpose of this section is to show how to combine the contractor method and the Schauder [1] fixed point principle.

Let W be a closed bounded convex set of a Banach space X .

Given two operators acting in X, $P:D(P) \to X$ and $Q:W \to X$, the following theorem is valid.

Theorem 7.1. Suppose that P is a closed operator having a bounded regular contractor Γ satisfying the inequality

$$(7.1) \qquad \|P(x+\Gamma(x)y) - Px - y\| \le q\|y\| \quad \text{with} \quad 0 < q < 1$$

for all $x \in D(P)$, $y \in X$, where $D(P) = \Gamma(x)(Y)$ and $\|\Gamma(x)\| \le B$. Suppose that Q is completely continuous and

$$(7.2) \qquad Qx - Py \in W \quad \text{and} \quad x + \Gamma(x)y \in W$$

for arbitrary $x,y \in W$. Then P and Q have a coincidence point $x^* \in W$, i.e., $Px^* = Qx^*$.

Proof. By virtue of Theorem 5.1, for arbitrary $x \in W$, the equation $Py = Qx$ has a unique solution y which is a limit of the iteration procedure $x_{n+1} = x_n - \Gamma(x_n)[Px_n - Qx]$. It follows from (7.2) that $x_n \in W$ for $n = 0,1,2,\cdots$. Hence, $y \in W$ and $y = P^{-1}Qx$ is completely continuous, since the inverse P^{-1} exists and is continuous, by Theorem 5.1. The Schauder fixed point theorem yields the existence of a point $x^* \in W$ such that $x^* = P^{-1}Qx^*$, i.e., $Px^* = Qx^*$.

Consider now the case where $Px = x - Fx$.

Theorem 7.2. Suppose that $F:D(F) \to X$, is a closed operator having a bounded regular contractor $\{I + \Gamma(x)\}$ such that the inequality $\|F(x+\Gamma(x)y) - F(x) - \Gamma(x)y\| \le q\|y\|$, $0 < q < 1$ is satisfied for all $x \in D(F)$ and $y \in X$, where $\|\Gamma(x)\| \le B$ and $D(F) = \Gamma(x)(Y)$. Suppose that Q is completely continuous and

$$(7.3) \qquad Fy+Qx \in W \quad \text{and} \quad x+\Gamma(x)[y-x] \in W \quad \text{for arbitrary} \quad x,y \in W.$$

Then there exists a fixed point $x^* \in W$ such that $x^* = Fx^* + Qx^*$.

Proof. By virtue of Theorem 5.2 and Remark 5.2, for arbitrary $x \in W$, the equation $y - Fy = Qx$ has a unique solution y which is a limit of the iteration procedure $x_{n+1} = x_n - [I+\Gamma(x_n)][x_n-Fx_n-Qx]$. Conditions (7.2) imply that $x_n \in W$ for $n = 0,1,2,\cdots$ and, consequently, $y \in W$. It follows from the existence and continuity of the inverse P^{-1} that $P^{-1}Qx$ is completely continuous, where $Px = x - Fx$. Applying Schauder's fixed point theorem, we obtain the existence of $x^* \in W$ such that $x^*-Fx^* = Qx^*$.

Both Theorems 7.1 and 7.2 can be considered as a generalization of the following theorem of Krasnosel'skii [2]:

If F is a contraction (i.e., Lipschitz continuous with a constant $q < 1$) and Q is completely continuous and $Py + Qx \in W$ for arbitrary $x,y \in W$, then there exists a fixed point $x^* \in W$ such that $x^* = Fx^* + Qx^*$. If the assumptions of Krasnosel'skii's theorem are satisfied, then by putting in Theorem 7.2, $\Gamma(x) \equiv 0$, the identity mapping I will be a bounded regular contractor satisfying conditions (7.2).

8. Stationary points of nonlinear functionals

Let $F:X \to R$ be a nonlinear functional differentiable in some ball $S(x_0,r)$ of X and denote by $F'(x)$ the Fréchet derivative (gradient) of F at x . The problem of unconstraint optimization of Fx reduces practically to finding stationary points x of F , i.e., satisfying the equation

(8.1) $$F'x = 0 , \quad x \in X .$$

Considering $F':X \to X'$ as a nonlinear (gradient) operator from the Banach space X into its dual Banach space X' , we can apply the contractor method to solve equation (8.1).

<u>Theorem 8.1</u>. Suppose that the gradient F' has a contractor $\Gamma(x):X' \to X$ for every $x \in S(x_0,r)$ such that

(8.2) $\|F'(x+\Gamma(x)y) - F'x - y\| \leq q\|y\|$ for $x \in S(x_0,r)$, $y \in X'$, $\|y\| \leq \eta$, where $\|\Gamma(x)\| \leq B$, $\|F'x_0\| \leq \eta$ and $B\eta(1-q)^{-1} \leq r$ for some constants $0 < q < 1$, B and η . If F' is closed on $S(x_0,r)$, then it has a stationary point $x^* \in S(x_0,r)$, i.e., $F'x^* = 0$, which is a limit of the sequence $\{x_n\}$ defined as follows $x_{n+1} = x_n - \Gamma(x_n)F'x_n$, $n = 0,1,2,\cdots$ and the error estimate is given by the formula $\|x_n - x^*\| \leq$ $\leq B\,\eta q^n(1-q)^{-1}$.

Proof. The proof is exactly the same as that of Theorem 2.1.

Remark 8.1. If the contractor is, additionally, onto and satisfies (8.2) for all $y \in X'$ such that $\|\Gamma(x)y\| \leq 2r$ for $x \in S(x_0,r)$, then F has in $S(x_0,r)$ a unique stationary point x^* .

This assertion follows from the argument in Remark 2.1.

Remark 8.2. It is sufficient to assume that F' is defined on a subset $D(F') = S(x_0,r) \cap \Gamma(x)(X')$, where the intersection is the same for all $x \in D(F')$ and $x_0 \in D(F')$. In this case, the uniqueness for x^* follows if condition (8.2) holds for all $y \in X'$ such that $\|\Gamma(x)y\| \leq 2r$, $x \in D(F')$.

9. Various iteration procedures as special cases of the contractor method

The Newton-Kantorovič method

It is shown in Section 2, that the Newton-Kantorovič method (2.11) is a special case of the contractor method and the same is true for the generalization of Newton's method (2.13) for nonlinear functionals. The following modification of the Newton-Kantorovič method is given by Bartle [1].

$$x_{n+1} = x_n - [P'(z_n)]^{-1}Px_n \,, \tag{9.1}$$

provided that the initial guess x_0 and the arbitrarily selected points z_n are sufficiently close to the solution desired. It is easy to see that the contractor for the method (9.1) is $\Gamma(x) = [P'(z)]^{-1}$. Under the assumptions made by Bartle [1], the contractor satisfies the following inequality,

$$\|P(x+[P'(z)]^{-1}y) - Px - y\| \leq (1/2\lambda)\|[P'(z)]^{-1}y\| \leq (1/2\lambda)\lambda\|y\| ,$$

where λ and β are chosen so as to satisfy $\|[P'(z)]^{-1}\| < \lambda$ if $\|z-x_0\| \leq \beta$, $\|x-x_0\| \leq \beta$ and $\|x+[P'(z)]^{-1}y-x_0\| \leq \beta$ (see Lemma 2, Bartle [1]).

Generalized gradient methods

It is shown by Altman [20] that the method of steepest descent developed by Kantorovič [1], the minimum residual method investigated by Krasnosel'skii and Krein [1] and other gradient methods (see Altman [21], Kivistik [1], Krasnosel'skii [2]) are special cases of Newton's method (2.13) for nonlinear functionals. Thus, these methods can also be considered as special cases of the contractor method. This is the case from variational point of view, i.e., when we reduce the operator equation to the minimum problem of a non-negative nonlinear functional F , where for instance, $F(x) = \|Px\|^2 = 0$ is required. However, we can see that these and other methods can also be considered as contractor methods in the direct sense of Section 3.

The minimum residual method

Let $A: H \to H$ be a linear self-adjoint and positive definite operator in the real Hilbert space H such that $m(x,x) \leq (Ax,x) \leq M(x,x)$, where $0 < m < M < \infty$. Consider the equation (9.2)

$$Ax = b , \quad x,b \in H . \tag{9.2}$$

The operator $Px = Ax - b$ is differentiable in the sense of Fréchet and $P'(x) = A$. It is easy to verify that $\|\alpha A - I\| < 1$ if $0 < \alpha < 2/M$. Thus, putting in (3.1^*), $\Gamma(x) = \alpha I$, we obtain a contractor and the corresponding contractor method will be the method of successive approximation with parameter α :

$$x_{n+1} = x_n - \alpha(Ax_n - b) , \quad n = 0,1,2,\cdots . \tag{9.3}$$

Replacing in (9.3), α by $\alpha(x) = \frac{(r,Ar)}{(Ar,Ar)}$, where $r = r(x) = Ax - b$, we obtain the

minimum residual method investigated by Krasnosel'skii and Krein [1]:

$$x_{n+1} = x_n - \alpha_n r_n \ , \quad n = 0,1,2,\cdots \ , \tag{9.4}$$

where $\alpha_n = \alpha(x_n)$ and $r_n = r(x_n)$. The contractor here is $\alpha(x)I$ and the contractor inequality (3.1) yields in this case

$$\|A(x+\alpha(x)y - Ax - y\| \le (M-m)(M+m)^{-1}\|y\| \ .$$

This inequality is satisfied for $y = r(x)$, by virtue of the following inequality (see Krasnosel'skii [2], p. 109),

$$\|A(x-\alpha(x)r(x)) - b\| \le (M+m)(M-m)^{-1}\|r(x)\| \ . \tag{9.5}$$

To prove the last inequality, let us observe that

$$\|A(x-\alpha(x)r(x)) - b\|^2 = \min_t \|A(x-tr(x) - b\|^2 = \min_t \|r(x) - tAr(x)\|^2 \ .$$

Bur for $t = 2(M+m)^{-1}$ we have $\|I - \frac{2}{M+m} A\| = (M-m)(M+m)^{-1}$ and, consequently, we obtain (9.5).

The method of steepest descent

The method of steepest descent developed by Kantorovič [1] for solving (9.2) is defined as follows,

$$x_{n+1} = x_n - \beta_n r_n \ , \tag{9.6}$$

where $\beta_n = (r_n, r_n)(Ar_n, r_n)^{-1}$. Consider the Hilbert space H_A obtained from H by introducing a new scalar product $[u,v] = (A^{-1}u,v)$, $u,v \in H$. Then it is clear that the steepest descent method (9.6) is the minimum residual method (9.4) considered in the Hilbert space H_A . Thus, the steepest descent method (9.6) is a contractor method in the Hilbert space H_A . It is easily seen that A is self-adjoint and positive definite in H_A .

A generalized gradient method in Hilbert space

Consider the nonlinear operator equation,

$$Px = 0 , \tag{9.7}$$

where $P:S(x_0,r) \to H$, P being continuously differentiable in the sense of Fréchet in the ball $S(x_0,r) \subset H$ and $P'(x)$ satisfies the inequality

$$\|P'(x)y\| \geq B^{-1}\|y\| \quad \text{for all} \quad x \in S(x_0,r) , \quad y \in H . \tag{9.8}$$

The following iteration procedure (see Altman [19]) is also a contractor method,

$$x_{n+1} = x_n - \|Px_n\|^2 \|Q(x_n)\|^{-2} 2^{-1} Q(x_n) , \quad n = 0,1,\cdots , \tag{9.9}$$

where $Q(x) = [P'(x)]^* Px$ (* = adjoint). Since P is differentiable, we can expect that $\Gamma(x) = \|Px\|^2 \|Q(x)\|^{-2} 2^{-1} [P'(x)]^*$ will be a contractor, by virtue of (3.1^*) . Condition (9.8) implies that $\|\Gamma(x)\|$ is bounded for $x \in S(x_0,r)$. The existence of a solution of (9.7) as well as the convergence of (9.9) to this solution can be obtained from Theorem 3.1. It follows from the assumptions made by Altman [19], that Theorem 3.1 can be applied. It is not difficult to see that the hypotheses made by Kivistik (see Krasnosel'skii [2], p. 156) are also sufficient in order to apply the contractor method to his procedure

$$x_{n+1} = x_n - (P'(x_n)Px_n, Px_n) \|P'(x_n)Px_n\|^{-2} Px_n , \quad n = 0,1,2,\cdots .$$

Note that other procedures similar to (9.9) can also be put in the unified scheme of the contractor method.

Remark 9.1. In all theorems which are global in nature, the requirement that the contractor $\Gamma(x)$ is bounded, i.e., $\|\Gamma(x)\| \leq B$ for $x \in D(P)$, can be replaced by the assumption that $\Gamma(x)$ is Lipschitz continuous with constant K .

Proof. We have, by (2.2),

$$\|\Gamma(x_{n+1})\| \leq \|\Gamma(x_n)\|(1+K\|Px_n\|) \leq \|\Gamma(x_n)\|(1+Kq^n\|Px_0\|) \leq \|\Gamma(x_n)\|(1+\tau)$$

for large n .

Hence,

$$\|\Gamma(x_{n+1})\|\|Px_{n+1}\| \leq (1+\tau)q\|\Gamma(x_n)\|\|Px_n\| ,$$

where $\tau < q^{-1} - 1$. Then P^{-1} will be continuous but not Lipschitzian.

A generalized gradient method in L^p space

Let $f \in L^q$ be a linear continuous functional on $L^p = L^p(a,b)$, $p > 1$, $1/p + 1/q = 1$. Putting

$$g(t) = \text{sign } f(t)\,|f(t)|^{1/(p-1)} \quad (a \leq t \leq b)$$

we have

$$\int_a^b f(t)g(t)\,dt = \|f\|_{L^q}^q \quad \text{and} \quad g \in L^p .$$

Let $F(x)$ be a nonlinear continuously differentiable functional defined on L^p and denote by $F'(x)$ the Fréchet derivative. Then the method (2.13), where $X = L^p$, $\|h_n\| = 1$ and $F'(x_n)h_n = \|F'(x_n)\|$, can be written as follows (see Altman [21]).

$$x_{n+1} = x_n - F(x_n)\|F'(x_n)\|^{-q}g(x_n) , \quad n = 0,1,\cdots ,$$

where $g(x)(t) = \text{sign } F'(x)(t)\,|F'(x)(t)|^{1/(p-1)}$.

Remark 9.2. A case discussed by Ehrmann [1] can be considered as a particular one where the contractor is a fixed linear bounded operator.

CHAPTER 2

CONTRACTORS WITH NONLINEAR MAJORANT FUNCTIONS

Introduction. The concept of a contractor has been introduced as a tool for solving equations in Banach spaces. In this way, various existence theorems for solutions of equations are obtained as well as convergence theorems for a broad class of iterative procedures. The contractor method yields also a unified approach to a large variety of iterative processes different in nature. The concept of a majorant function is due to Kantorovič who used it for comparison of the general Newton method with the same for one dimension. In this chapter, both concepts are combined; the contractor concept and the concept of a majorant function. In this way, i.e., by using contractors with nonlinear majorant functions†, not only existence and convergence theorems can be obtained for a large class of iterative procedures but the character of the convergence itself can also be investigated. On this basis, a unified theory is developed for a very large variety of iterative methods including the most important ones.

An application of the method of contractors with nonlinear majorant functions to the Quantum Field Theory is given by Yoshimura [1].

2. General majorant functions and contractors

Let $P:D \subset X \to Y$ be a nonlinear operator, X and Y being Banach spaces and let $Q(t) \geq 0$, $t \geq 0$, $Q(0) = 0$, be a nondecreasing continuous function. Suppose that $\Gamma(x):Y \to X$ is a linear bounded operator associated with $x \in X$.

---

† Let us mention that in our case, the role and character of the majorant functions are rather different.

Definition 2.1. We say that $\Gamma(x)$ is a contractor for P with nonlinear majorant Q if for $x \in D$, $y \in Y$, the following contractor inequality is satisfied

(2.1) $$\|P(x+\Gamma(x)y) - Px - y\| \leq Q(\|y\|)$$

whenever $x+\Gamma(x)y \in D$; x and y are to be specified for the problem of solving the nonlinear equation,

(2.2) $$Px = 0 , \quad x \in D \subset X .$$

The contractor $\Gamma(x)$ is said to be bounded on S if there exists a constant B such that

(2.3) $$\|\Gamma(x)\| \leq B \quad \text{for all} \quad x \in S \subset D .$$

In order to solve (2.2), let us consider the following iterative procedures,

(2.4) $$x_{n+1} = x_n - \Gamma(x_n)Px_n , \quad n = 0,1,\cdots ,$$

(2.5) $$t_{n+1} = t_n + Q(t_n - t_{n-1}) , \quad t_0 = 0 , \quad n = 1,2,\cdots ,$$

and assume that the sequence $\{t_n\}$ converges toward some $t^* > 0$. Denote by $S = S(x_0,r) \subset X$ the open ball with center x_0, and radius $r = Bt^*$.

The operator P is said to be closed if $x_n \in D$, $x_n \to x$ and $Px_n \to y$ imply $x \in D$ and $y = Px$ or the graph of P is closed.

Theorem 2.1. Suppose that $P: D \subset X \to Y$ is a closed operator. Let $x_0 \in D$ be chosen so as to satisfy

(2.6) $$S(x_0, Bt^*) \subset D \quad \text{and} \quad \|Px_0\| \leq t_1 .$$

If P has a bounded contractor Γ satisfying (2.3) and (2.1), then all x_n lie in S and the sequence $\{x_n\}$ defined by (2.4) converges toward a solution x of equation (2.2) and the error estimate (2.7) holds,

(2.7) $$\|x-x_n\| \leq t^* - t_n , \quad n = 0,1,\cdots .$$

<u>Proof</u>. Since $Q(t) \geq 0$ for $t \geq 0$, it follows from (2.5), that

$$0 = t_0 \leq t_1 \leq t_2 \leq \cdots \leq t^* .$$

We have, by (2.3), (2.4) and (2.6),

$$\|x_1-x_0\| \leq B\|Px_0\| \leq Bt_1 \leq Bt^* .$$

Using the contractor inequality (2.1) with $x = x_{n-1}$ and $y = -Px_{n-1}$, we show by induction that

(2.8) $$\|Px_n\| \leq Q(\|Px_{n-1}\|) \leq Q(t_n-t_{n-1}) = t_{n+1} - t_n .$$

Hence, we conclude from (2.3) - (2.5) that

(2.9) $$\|x_{n+1}-x_n\| \leq B(t_{n+1}-t_n) ,$$

(2.10) $$\|x_n-x_0\| \leq Bt_n \leq Bt^* .$$

Inequality (2.10) shows that $x_n \in S(x_0,Bt^*)$. Using a standard argument, we derive from (2.9) that the sequence $\{x_n\}$ converges to some element $x \in \bar{S}$ (closure of S), since $Px_n \to 0$, by (2.8), and P is closed, it follows that $Px = 0$.

The error estimate (2.7) results from (2.9) in the usual way.

In order to obtain a fixed point theorem for $F:X \to X$, we put

$$Px = x-Fx \quad \text{and} \quad Y = X .$$

Then the contractor inequality (2.1) becomes

(2.11) $$\|Fx - F(x+\Gamma(x)y) - (I - \Gamma(x))y\| \leq Q(\|y\|) ,$$

where I is the identity mapping of X. In the special case, where $\Gamma(x) \equiv I$, for all corresponding $x \in X$, the contractor inequality (2.11) yields

(2.12) $$\|F(x+y) - Fx\| \leq Q(\|y\|) .$$

If we substitute in (2.12), $y = Fx - x$, then we obtain the following contractor inequality,

$$\|F^2x - Fx\| \leq Q(\|Fx-x\|) \tag{2.13}$$

and this is exactly the case systematically investigated by Ortega and Rheinboldt [2], 12.4. In this case, (2.4) yields the simple successive approximation method,

$$x_{n+1} = Fx_n , \quad n = 0,1,\cdots .$$

Thus, the successive approximation method with the nonlinear majorant satisfying (2.13) is a special case of the general contractor method with the nonlinear majorant satisfying the contractor inequality (2.1).

Remark 2.1. Suppose that the contractor $\Gamma(x)$ is a mapping onto or has the following property. For arbitrary solutions $x \in S$ and $\bar{x} \in S$ of (2.2), there exists an element y of Y such that

$$x + \Gamma(x)y = \bar{x} . \tag{2.14}$$

Then (2.2) has a unique solution if $Q(t) < t$ for $t > 0$.

Proof. It follows from (2.1) that $\|y\| \leq Q(\|y\|)$ and we obtain a contradiction by assuming $\|y\| > 0$.

Notice that (2.14) is satisfied in the special case just mentioned, that is, in the case of the successive approximation method with nonlinear majorant subject to (2.12). Then $\Gamma(x) \equiv I$, $Y = X$ and we put, obviously, $y = \bar{x} - x$ in (2.14).

Remark 2.2. Putting in (2.1), $Q(t) = qt$ with $0 < q < 1$, we obtain the following contractor inequality,

$$\|P(x+\Gamma(x)y) - Px - y\| \leq q\|y\| .$$

In this case, $\Gamma(x)$ is a contractor with a linear majorant Q. This is precisely the case investigated in Chapter 1.

Now let us consider the case of a contractor with more general linear majorant $Q(t) = qr + \gamma$, where $0 < q < 1$ and $\gamma \geq 0$. Thus, the contractor inequality is

$$\|P(x+\Gamma(x)y) - Px - y\| \leq q\|y\| + \gamma \ . \tag{2.15}$$

Theorem 2.2. Let $P:D \subset X \to Y$ be a closed nonlinear operator. Suppose that $x_0 \in D$ is chosen so as to satisfy

$$S = S(x_0, Bt^*) \subset D \quad \text{and} \quad \|Px_0\| \leq \gamma ,$$

where $t^* = \gamma(1-q)^{-1}$. If P has a bounded linear contractor Γ satisfying (2.3) and the contractor inequality (2.15) with $x \in S$ and $\|y\| \leq t^*$, then all x_n lie in S and the sequence $\{x_n\}$ defined by (2.4) converges toward a solution x of equation (2.2) and the error estimate holds

$$\|x-x_n\| \leq \gamma q^n (1-q)^{-1} , \quad n = 0,1,\cdots .$$

Proof. The proof is exactly the same as that of Theorem 2.1 in which we put $Q(t) = qt + \gamma$.

3. Quadratic majorant functions and contractors

There is in the literature a variety of iterative methods which are closely connected with quadratic majorants. These methods and a discussion of them can be included in a general discussion of methods using contractors with quadratic majorants.

Let $P:D \subset X \to Y$ be a closed nonlinear operator having a contractor $\Gamma(x)$ with quadratic majorant,

$$Q(t) = at^2 + bt , \quad a > 0 , \quad b \geq 0 ,$$

that is, satisfying the contractor inequality,

$$\|P(x+\Gamma(x)y) - Px - y\| \leq a\|y\|^2 + b\|y\| , \tag{3.1}$$

for $x \in D$, whenever $x + \Gamma(x)y \in D$. The corresponding majorant iteration procedure is

(3.2) $$t_{n+1} = Q(t_n) , \quad t_0 = \eta , \quad n = 1,2,\cdots .$$

Theorem 3.1. Suppose that $x_0 \in D$ is chosen so as to satisfy

(3.3) $$S = S(x_0, Bt^*) \subset D \quad \text{and} \quad \|Px_0\| \le \eta ,$$

(3.4) $$q = a\eta + b < 1 , \quad t^* = \eta(1-q)^{-1} .$$

If P has a bounded linear contractor $\Gamma(x)$ satisfying (2.3) and the contractor inequality (3.1) for $x \in S \subset D$, then all x_n lie in S and the sequence $\{x_n\}$ defined by (2.4) converges to a solution x of equation (2.2) and the error estimate holds,

(3.5) $$\|x-x_n\| \le B\,\eta q^n(1-q)^{-1} , \quad n = 0,1,\cdots .$$

Proof. First, note that it follows from (2.4) and (3.1) with $x = x_n$ and $y = -Px_n$ that

$$\|Px_{n+1}\| \le Q(\|Px_n\|) , \quad n = 0,1,\cdots .$$

Further, by virtue of (3.2) and (3.3), we prove by induction that

(3.6) $$\|Px_n\| \le t_n , \quad n = 0,1,2,\cdots .$$

Now we prove also by induction that

(3.7) $$t_{n+1} \le qt_n \le q^{n+1}\eta \le \eta , \quad n = 0,1,\cdots .$$

In fact, we have, by (3.2) and (3.4), $t_1 = (at_0+b)t_0 = (a\eta+b)t_0 = qt_0$. Using the assumption for the induction, we obtain $t_{n+1} = (at_n+b)t_n \le$ $\le (a\eta+b)t_n \le qt_n$. In the standard way, we derive from (2.4), (3.6) and (3.7) that if $m > n$, then

(3.8) $$\|x_m-x_n\| \le Bq^n\eta(1-q^{m-n-1})(1-q)^{-1} ,$$

and all assertions of the theorem result from (3.8), since $Px_n \to 0$ as $n \to \infty$, by (3.6) and (3.7).

An application of Theorem 3.1 can be given by using the following.

Lemma 3.1. Let $P:D \subset X \to Y$ be a nonlinear operator differentiable in the sense of Fréchet and let its derivative $P'(x)$ be Lipschitz continuous on some ball $S = S(x_0,r) \subset D$, i.e., there exists a constant K such that

$$\|P'(x) - P'(\bar{x})\| \le K\|x-\bar{x}\| , \tag{3.9}$$

for $x,\bar{x} \in S$. Let $A(x):X \to Y$ be a bounded linear nonsingular operator such that

$$\|A(x)^{-1}\| \le B \quad \text{and} \quad \|P'(x) - A(x)\| \le C \tag{3.10}$$

for $x \in S$ and some constants B, C. Then $\Gamma(x) = A(x)^{-1}$ is a bounded linear contractor for P with quadratic majorant
$Q(t) = \frac{1}{2}B^2Kt^2 + C\,Bt$, i.e.,

$$\|P(x+\Gamma(x)y) - Px - y\| \le \frac{1}{2}KB^2\|y\|^2 + C\,B\|y\|$$

for $x \in S$, whenever $x+\Gamma(x)y \in S$.

Proof. We have

$$\begin{aligned}\|P(x+\Gamma(x)y) - Px - y\| &\le \|P(x+\Gamma(x)y) - Px - P'(x)\Gamma(x)y\| + \\ &+ \|P'(x)\Gamma(x)y - A(x)\Gamma(x)y\| \le \frac{1}{2}K\|\Gamma(x)y\|^2 + C\|\Gamma(x)y\| \le \\ &\le \frac{1}{2}KB^2\|y\|^2 + CB\|y\| .\end{aligned}$$

Theorem 3.2. Suppose that $P:D \subset X \to Y$ satisfies the hypotheses of Lemma 3.1. If, in addition,

$$\|Px_0\| \le \eta , \quad q = \frac{1}{2}B^2K\eta + BC < 1 , \quad r = Bt^* \quad \text{and} \quad t^* = \eta(1-q)^{-1} ,$$

then the sequence $\{x_n\}$ defined by the iteration process,

$$x_{n+1} = x_n - A(x_n)^{-1}Px_n , \quad n = 0,1,\cdots$$

converges to a solution x of equation (2.2); all x_n lie in S and the error estimate (3.5) holds.

Proof. By Lemma 3.1, P has a bounded linear contractor $\Gamma(x) = A(x)^{-1}$ satisfying the assumptions of Theorem 3.1 with $a = \frac{1}{B}B^2K$ and $b = BC$. Since all hypotheses of Theorem 3.1 are fulfilled, the proof follows immediately.

Lemma 3.2. Let $P:D \subset X \to Y$ be a closed nonlinear operator and let $T:D \subset X \to Y$ be differentiable in the sense of Fréchet on $S = S(x_0,r) \subset D$. Moreover, suppose that $\|T'(x) - T'(\bar{x})\| \leq K\|x-\bar{x}\|$ and $\|(Px-Tx) - (P\bar{x}-T\bar{x})\| \leq C\|x-\bar{x}\|$ for all $x,\bar{x} \in S$, and $T'(x)$ is nonsingular and $\|T'(x)^{-1}\| \leq B$ for all $x \in S$. Then $\Gamma(x) = T'(x)^{-1}$ is a bounded linear contractor for P satisfying the contractor inequality

$$\|P(x+\Gamma(x)y) - Px - y\| \leq \frac{1}{2}KB^2\|y\|^2 + CB\|y\|$$

for all $x \in S$ whenever $x+\Gamma(x)y \in S$.

Proof. We have

$$\|P(x+\Gamma(x)y) - Px - y\| \leq \|T(x+\Gamma(x)y) - Tx - T'(x)\Gamma(x)y\| +$$

$$+ \|[P(x+\Gamma(x)y) - T(x+\Gamma(x)y)] - [Px-Tx]\| \leq$$

$$\leq \frac{1}{2}K\|\Gamma(x)y\|^2 + C\|\Gamma(x)y\| \leq \frac{1}{2}KB^2\|y\|^2 + CB\|y\|$$

for all $x \in S$ whenever $x+\Gamma(x)y \in S$.

By using Lemma 3.2, we obtain as a special case of Theorem 3.1, the following theorem which is due to Zinčenko [1].

Theorem 3.3. Suppose that, in addition to the hypotheses of Lemma 3.2, the following conditions are satisfied:

$$\|Px_0\| \leq \eta, \quad q = \frac{1}{2}B^2K\eta + BC < 1, \quad r = Bt^* \text{ and } t^* = \eta(1-q)^{-1}.$$

Then the sequence $\{x_n\}$ defined by the iteration procedure

$$x_{n+1} = x_n - T'(x)^{-1} Px_n \ , \quad n = 0,1,\cdots$$

converges to a solution of equation $Px = 0$. All x_n lie in S and the error estimate holds,

$$\|x-x_n\| \leq B\,\eta q^n (1-q)^{-1} \ , \quad n = 0,1,\cdots \ .$$

Proof. By Lemma 3.2, P has a bounded linear contractor $\Gamma(x) = T'(x)^{-1}$ satisfying the assumptions of Theorem 3.1 with

$$a = \frac{1}{2} B^2 K \quad \text{and} \quad b = BC \ .$$

Since all hypotheses of Theorem 3.1 are fulfilled, the proof follows immediately.

Let us consider a special case of a contractor with quadratic majorant $Q(t) = at^2$, i.e., the contractor inequality is

$$\|P(x+\Gamma(x)y) - Px - y\| \leq a\|y\|^2 \ . \tag{3.11}$$

Theorem 3.4. Let $P: D \subset X \to Y$ be a closed nonlinear operator with domain D containing the ball $S = S(x_0, r)$, where $\|Px_0\| \leq \eta$, $q = a\eta < 1$, $r = B\,\eta t^*$ and $t^* = \sum_{n=0}^{\infty} q^{2^n-1}$. Suppose that P has a bounded linear contractor $\Gamma(x)$ satisfying (2.3) and the contractor inequality (3.11) for $x \in S$ whenever $x+\Gamma(x)y \in S$. Then the sequence $\{x_n\}$ defined by (2.4) converges to a solution x of the equation $Px = 0$. All x_n lie in S and the error estimate holds,

$$\|x-x_n\| \leq B\,\eta \sum_{i=n}^{\infty} q^{2^i-1} \leq B\,\eta t^* q^{2^n-1} \ . \tag{3.12}$$

Proof. By using the induction, it follows from (3.11) with $x = x_n$ and $y = -Px_n$ that

$$\|Px_n\| \leq a\|Px_{n-1}\|^2 \leq a\,t_{n-1}^2 = t_n \ , \tag{3.13}$$

where $t_{n+1} = a\,t_n^2$, $t_0 = \eta$, $n = 0,1,\cdots$. We prove, by induction, that

(3.14) $$t_n = \eta q^{2^n-1} , \quad n = 0,1,\cdots .$$

Hence, by virtue of (2.4), (3.13) and (3.14),

$$\|x_{n+1}-x_n\| \leq B \eta q^{2^n-1} ,$$

$$\|x_{n+1}-x_0\| \leq B \eta \sum_{i=0}^{n} q^{2^i-1} < B \eta t^* .$$

The latter inequality shows that all x_n lie in S . Further, we have

(3.15) $$\|x_{n+p}-x_n\| \leq B \eta \sum_{i=n}^{n+p-1} q^{2^i-1} < B \eta t^* q^{2^n-1} .$$

Thus, $\{x_n\}$ is a Cauchy sequence and has a limit element x . It follows from (3.13) and (3.14) that $Px_n \to 0$ as $n \to \infty$. Since P is closed, we conclude that $Px = 0$. The error estimate follows from (3.15).

The following theorem is due to Mysovskih [1] and is also presented in Kantorovič and Akilov [1] (see also Ortega and Rheinboldt [2]). This is a special case of Theorem 3.4 where $\Gamma(x) = P'(x)^{-1}$.

Theorem 3.5. Suppose that $P:D \subset X \to Y$ is differentiable in the sense of Fréchet in the ball $S = S(x_0,r) \subset D$ and $P'(x)$ is nonsingular in S and satisfies the following conditions,

(3.16) $$\|P'(x) - P'(\bar{x})\| \leq K\|x-\bar{x}\| \quad \text{for} \quad x,\bar{x} \in S ,$$

(3.17) $$\|P'(x)^{-1}\| \leq B \quad \text{for} \quad x \in S(x_0,r) ,$$

where $r = B \eta t^*$, $t^* = \sum_{n=0}^{\infty} a^{2^n-1}$, $q = \frac{1}{2} B^2 K \eta < 1$ and $\|Px_0\| \leq \eta$.

Then the sequence $\{x_n\}$, defined by Newton's method

$$x_{n+1} = x_n - P'(x_n)^{-1} Px_n , \quad n = 0,1,\cdots ,$$

converges to a solution x of equation $Px = 0$. All x_n lie in S and the error estimate (3.12) holds.

Proof. We have

$$\|P(x+P'(x)^{-1}y) - Px - y\| = \|P(x+P'(x)^{-1}y) - Px - P'(x)P'(x)^{-1}y\| \leq$$

$$\leq \frac{1}{2} K \|P'(x)^{-1}y\|^2 \leq \frac{1}{2} KB^2 \|y\|^2 ,$$

by (3.16) and (3.17). Thus, P has a bounded linear contractor $\Gamma(x) = P'(x)^{-1}$ satisfying (3.11) with $a = \frac{1}{2} B^2 K$. It is easily seen that all hypotheses of Theorem 3.4 are fulfilled and the proof is complete.

Remark 3.1. The contractor inequality (3.11) implies that the contractor $\Gamma(x)$ is an inverse Fréchet derivative (see Chapter 1). Thus, $\Gamma(x)$ is a one-to-one mapping, moreover, $\Gamma(x)$ is a homeomorphism onto a subspace of X. This assertion follows from the inequality

$$\|y\|(1-a\|y\|) \leq \|P(x+\Gamma(x)y) - Px\|$$

for sufficiently small $\|y\|$. This inequality is an immediate consequence of (3.11). Our argument requires the continuity of P.

4. Non-bounded contractors

So far, we discussed only such cases where the contractors were bounded by constants. In this section, we consider a more general case.

Let $P: D \subset X \to Y$ be a nonlinear operator and let $x_0 \in D$ be chosen so that $S = S(x_0, r) \subset D$, where the radius r will be defined below. Let $\Gamma(x): Y \to X$ be a bounded linear operator associated with $x \in D$. We assume that $\Gamma(x)$ is a contractor for P with majorant function Q satisfying the following contractor inequality,

$$\|P(x+\Gamma(x)y) - Px - y\| \leq Q(\|y\|, \|x-x_0\|) \tag{4.1}$$

for $x \in S$ and $y \in Y$ whenever $x+\Gamma(x)y \in D$, $Q(w,t) \geq 0$ for $w,t \geq 0$ being continuous and nondecreasing in each variable. Suppose that the following estimate holds for the contractor Γ,

$$\|\Gamma(x)\| \leq B(\|x-x_0\|) \quad \text{for} \quad x \in S, \tag{4.2}$$

where $B(t) \geq 0$ for $t \geq 0$ is a nondecreasing function and $B(0) > 0$.

In order to investigate the iterative process (2.4), we introduce the following two numerical iterative sequences,

$$t_{n+1} = t_n + Q(t_n - t_{n-1}, s_n), \quad n = 1,2,\cdots \tag{4.3}$$

with initial values $t_0 = 0$, $t_1 = Q(\eta, 0)$,

$$s_{n+1} = s_n + B(s_n)(t_n - t_{n-1}), \quad n = 1,2,\cdots \tag{4.4}$$

with initial values $s_0 = 0$, $s_1 = B(0)\eta$. Hence, we obtain from (4.3) and (4.4),

$$s_{n+1} = s_n + B(s_n) Q((s_n - s_{n-1})/B(s_{n-1}), s_{n-1}), \quad n = 1,2,\cdots. \tag{4.4a}$$

We assume that the sequence $\{s_n\}$ is convergent and $s^* = \lim_{n\to\infty} s_n$.

<u>Theorem 4.1</u>. Suppose that $P: D \subset X \to Y$ is a closed nonlinear operator and $x_0 \in X$ is such that $S = S(x_0, r) \subset D$ and

$$\|Px_0\| \leq \eta, \quad B(0)\eta = s_1, \quad r = s^* \text{ and } Q(0, s^*) = 0. \tag{4.5}$$

Let $\Gamma: S \to L(Y \to X)$ be a contractor for P satisfying (4.2) and the contractor inequality (4.1). Then all x_n lie in S and the sequence $\{x_n\}$ defined by (2.4) converges to a solution x of $Px = 0$ and the error estimate holds,

$$\|x - x_n\| \leq s^* - s_n, \quad n = 0,1,\cdots. \tag{4.6}$$

<u>Proof</u>. Using (4.1) with $x = x_n$ and $y = -Px_n$, we show, by induction, that

$$\begin{aligned} \|Px_{n+1}\| &\leq Q(\|Px_n\|, \|x_n - x_0\|) \leq Q(t_n - t_{n-1}, s_n) = \\ &= t_{n+1} - t_n, \quad n = 1,2,\cdots, \end{aligned} \tag{4.7}$$

by virtue of (4.3) and (4.5), and

$$\|x_{n+1} - x_n\| \leq B(s_n)(t_n - t_{n-1}) = s_{n+1} - s_n, \quad n = 1,2,\cdots,$$

by virtue of (2.4), (4.2) and (4.4). Hence, it follows, by induction, that

(4.8) $\|x_n - x_m\| \leq s_n - s_m$ and $\|x_n - x_0\| \leq s_n$, $n > m = 0,1,\cdots$.

Thus, the sequence $\{x_n\}$ converges to some element x . Since $\{s_n\}$ is convergent toward s^* by assumption, it results from (4.4a), that

$$t_n - t_{n-1} = (s_{n+1} - s_n)/B(s_n) \to 0 \quad \text{as} \quad n \to \infty$$

and consequently $Px_n \to 0$ as $n \to \infty$ by virtue of (4.7) and (4.5). Since P is closed, we conclude that $Px = 0$. The error estimate (4.6) results from (4.8) by letting $n \to \infty$ and the proof is complete.

In order to discuss special cases of Theorem 4.1, we need some elementary lemmas concerning Newton-type iterative methods in one dimensional space. These methods are applied to real valued functions considered as majorants for contractors. Although these lemmas correspond to the case of existence of "first integrals" for nonlinear difference equations as called by Rheinboldt [1], the application of the contractor methods admits also more general cases. This follows, for instance, from Theorem 4.1.

The following simple lemma deals with the Newton-related mathod in one dimensional case.

Lemma 4.1. Let $u(t)$ be a real valued function continuous on $[0,t^*]$, where t^* is the smallest positive root of $u(t) = 0$. Let $v(t) > 0$ for $0 \leq t < t^*$ be continuous on $[0,t^*]$. Suppose that

(4.9) $$u(s) - u(t) + v(t)(s-t) > 0$$

for arbitrary $0 \leq t < s \leq t^*$. Then all $t_n < t^*$ and the sequence $\{t_n\}$ defined by

(4.10) $t_0 = 0$, $t_{n+1} = t_n + u(t_n)/v(t_n)$, $n = 0,1,\cdots$

with initial values $t_0 = 0$, $t_1 = \eta < t^*$ converges toward t^* and satisfies the difference equations

$$(4.11) \qquad t_{n+1} - t_n = [1/v(t_n)][u(t_n) - u(t_{n-1}) + v(t_{n-1})(t_n - t_{n-1})] \; .$$

Proof. (4.11) follows immediately from (4.10) and shows that the sequence $\{t_n\}$ is increasing. Set $g(t) = t+u(t)/v(t)$; then it follows from (4.9), that

$$(4.12) \qquad t^* - g(t) = [1/v(t)][u(t^*) - u(t) + v(t)(t^* - t)] > 0 \; .$$

By virtue of (4.12), it is easily seen by induction that

$$t_{n+1} = g(t_n) < t^* \quad \text{for} \quad n = 0,1,\cdots \; .$$

Hence, the sequence $\{t_n\}$ is bounded and, consequenctly, monotonically convergent to some number $t \leq t^*$. But $u(t_n)/v(t_n) \to 0$, by (4.10), yielding $u(t_n) \to 0$ as $n \to \infty$ and $u(t) = 0$. Hence, we conclude that $t = t^*$, since t^* is the smallest positive root of $u(t) = 0$.

As a special case of Lemma 4.1, we obtain the following.

Lemma 4.2. Let $u(t)$ be a real valued function with continuous second derivative $u''(t) > 0$ for $0 \leq t \leq t^*$, where t^* is the smallest positive root of $u(t) = 0$. Let $v(t) > 0$ for $0 \leq t < t^*$ be continuous on $[0,t^*]$. If $|u'(t)| \leq v(t)$ for $0 \leq t \leq t^*$, then all $t_n < t^*$ and the sequence $\{t_n\}$ defined by (4.10) converges toward t^* and satisfies (4.11).

Proof. It is easy to see that condition (4.9) is satisfied. In fact, we have for $0 \leq t < s \leq t^*$,

$$u(s) - u(t) + v(t)(s-t) = u(s) - u(t) - u'(t)(s-t) + [u'(t) + v(t)](s-t) > 0 \; ,$$

since $u'(t) + v(t) \geq 0$ and $u''(t) > 0$.

Lemma 4.3. Assume that $p_i \geq 0$, $i = 1,\cdots,4$, $p_1 > 0$ and $p_2 < 1$. Let $0 < \eta \leq (1-p_2)^2/4\bar{p}_1$, where $\bar{p}_1 = \max[p_1, (p_3+p_4)/2]$. Then the sequence $\{t_n\}$ defined by (4.10) with $u(t) = \bar{p}_1 t^2 - (1-p_2)t+\eta$ and $v(t) = 1 - p_4 t$ is strictly increasing and is the same as that defined by

$$t_{n+1} = t_n + [1/(1-p_4 t_n)][\bar{p}_1(t_n - t_{n-1}) + p_2 + (2\bar{p}_1 - p_4)t_{n-1}](t_n - t_{n-1})$$

with initial values $t_0 = 0$, $t_1 = \eta$ and

$$\lim_{n\to\infty} t_n = t^* = \{(1-p_2) - [(1-p_2)^2 - 4\bar{p}_1\eta]^{1/2}\}/2\bar{p}_1 .$$

The sequence $\{t_n\}$ majorizes the sequence $\{s_n\}$, that is, $s_{n+1} - s_n \le t_{n+1} - t_n$, $n = 0,1,\cdots$, where

$$s_{n+1} = s_n + [1/(1-p_4 s_n)][p_1(s_n - s_{n-1}) + p_2 + p_3 s_{n-1}](s_n - s_{n-1}) ,$$

with initial values $s_0 = 0$, $s_1 = \eta$. If $2p_1 = p_3 + p_4$, then $\{s_n\}$ and $\{t_n\}$ coincide.

Proof. Since

$$p_1 \le \bar{p}_1 \quad \text{and} \quad p_3 \le 2\bar{p}_1 - p_4 , \tag{4.13}$$

we have

$$u''(t) = 2\bar{p}_1 > 0 \quad \text{and} \quad u'(t) + v(t) = (2\bar{p}_1 - p_4)t + p_2 \ge 0 \quad \text{for} \quad t \ge 0 .$$

It remains to show that $v(t) > 0$ for $0 \le t < t^*$. In fact, we have

$$v(t) = 1 - p_4 t > 1 - p_4 t^* = (2\bar{p}_1)^{-1}\{2\bar{p}_1 - p_4 + p_4 p_2 + p_4[(1-p_2)^2 - 4\bar{p}_1\eta]^{1/2}\} \ge$$

$$\ge (2\bar{p}_1)^{-1}\{p_3 + p_4 p_2 + p_4[(1-p_2)^2 - 4\bar{p}_1\eta]^{1/2}\} > 0$$

for $t < t^*$ and $v(t^*) \ge 0$. Therefore, all hypotheses of Lemma 4.2 are satisfied. The majorization property results from (4.13) by induction.

Let us observe that Lemma 4.3 generalizes the following.

Lemma of Rheinboldt [1]. (See also Ortega and Rheinboldt [2], 12.6.3.)

Assume that $p_i > 0$, $i = 1,\cdots,4$, $p_1 > 0$, $p_2 < 1$ and

$p_3+p_4 = 2p_1$. Then, all $t_n < t^*$ and the sequence $\{t_n\}$ defined by

$$t_{n+1} = t_n + [1/(1-p_4 t)][p_1(t_n-t_{n-1}) + p_2 + p_3 t_{n-1}](t_n-t_{n-1}) \tag{4.14}$$
$$\text{for} \quad n = 1,2,\cdots$$

with initial values $t_0 = 0$, $t_1 = \eta$, $0 \leq \eta \leq (1-p_2)^2/4p_1$, is strictly increasing and $\lim_{n\to\infty} t_n = t^* = \{(1-p_2) - [(1-p_2)^2 - 4p_1\eta]^{1/2}\}/2p_1$.

Proof. Condition $p_3+p_4 = 2p_1$ implies $\bar{p}_1 = p_1$ and (4.14) coincides with (4.10).

The following lemma is due to Ortega and Rheinboldt [2], 12.6.2. Details of the proof given here are different.

Lemma 4.4. Let a and η be positive constants such that $h = a\eta \leq 1/2$. Then the sequence $\{t_n\}$ defined by

$$t_{n+1} = t_n + a(t_n-t_{n-1})^2/2(1-at_n) , \quad t_0 = 0 , \quad t_1 = \eta , \tag{4.15}$$

$n = 0,1,\cdots$ converges to the smaller root t^* of $(1/2)at^2 - t + \eta = 0$ and the error estimate holds,

$$t^* - t_n \leq (a2^n)^{-1}(2h)^{2^n} , \quad n = 0,1,\cdots . \tag{4.16$_n$}$$

The sequence defined by (4.15) is the same as that defined by

$$t_{n+1} = t_n + ((1/2)at_n^2 - t_n + \eta)/(1-at_n) \tag{4.17}$$

or by

$$t_{n+1} = ((1/2)at_n^2 - \eta)/(at_n - 1) . \tag{4.18}$$

Proof. Put $u(t) = (1/2)at^2 - t + \eta$ and $v(t) = 1 - at$. Then (4.10) and (4.17) coincide with (4.14) where $p_1 = a/2$, $p_2 = 0$, $p_3 = 0$ and $p_4 = a$, since $\bar{p}_1 = p_1$. It is easy to verify that if the sequence $\{t_n\}$ is defined by (4.17), then

$$(1/2)at_n^2 - t_n + \eta = (1/2)a(t_n-t_{n-1})^2 . \tag{4.19}$$

In fact, (4.19) is equivalent to

(4.20) $$t_n(1-at_{n-1}) = -(1/2)at_{n-1}^2 + \eta$$

and (4.20) follows from (4.17) with n replaced by $n-1$. Obviously, (4.18) results from (4.20). By Lemma 4.3, the sequence (4.17) converges to t^*. To prove the error estimate, we first show, by induction, that

(4.21) $$1/(1-at_n) \leq 2^n$$

or, equivalently,

(4.21_n) $$at_n \leq 1 - 1/2^n, \quad n = 0,1,\cdots .$$

It follows from (4.18), that

$$at_{n+1} = [a\eta - (at_n)^2/2]/(1-at_n) \leq [1/2 - (at_n)^2/2]/(1-at_n) =$$

$$= (1+at_n)/2 \leq (2-1/2^n)/2 = 1 - 1/2^{n+1},$$

by virtue of (4.21_n). Hence, (4.21_n) implies (4.21_{n+1}). For $n = 0$, condition (4.21_n) is obviously satisfied. Since $v(t) = -u'(t)$ and $u''(t) = 2\bar{p}_1 = 2p_1 = a$, it follows from (4.12) with $t = t_n$ that

$$t^*-t_{n+1} = a(t^*-t_n)^2/2(1-at_n) \leq 2^{n-1}a(t^*-t_n)^2 \leq (a2^{n+1})^{-1}(2h)^{2^{n+1}},$$

by virtue of (4.21) and (4.16_n). Thus, (4.16_n) implies (4.16_{n+1}) which holds obviously true for $n = 0$, since $1 - (1-2h)^{1/2} \leq 2h$.

As an application of Lemma 4.1, we obtain the following.

<u>Theorem 4.2</u>. Suppose that $P: D \subset X \to Y$ is a nonlinear closed operator and let Γ be a contractor for P satisfying (4.1) and (4.2). Assume that there exist two functions satisfying the hypotheses of Lemma 4.1 and such that

(4.22) $$B(s)Q((s-t)/B(0),t) \leq [1/v(s)][u(s) - u(t) + v(t)(s-t)]$$
$$\text{for } 0 \leq t < s \leq t^*, \quad n = 1,2,\cdots .$$

Finally, let $x_0 \in D$ be such that

(4.23) $$\|Px_0\| \leq \eta \,, \quad B(0)\eta < u(0)/v(0) \quad \text{and} \quad r = t^* = \lim_{n\to\infty} t_n \,,$$

where t^* is the smallest positive root of $u(t) = 0$ and $\{t_n\}$ is given by (4.10) or (4.11). Then all x_n lie in $S(x_0,r)$ and the sequence $\{x_n\}$ defined by (2.4) converges to a solution x of $Px = 0$ and the error estimate holds,

(4.24) $$\|x-x_n\| \leq t^* - t_n \,, \quad n = 0,1,\cdots \,.$$

<u>Proof</u>. It follows from (4.8) that

(4.25) $$\|x_{n+1}-x_n\| \leq s_{n+1}-s_n \quad \text{and} \quad \|x_n-x_0\| \leq s_n \,, \quad n = 0,1,\cdots \,,$$

where the sequence $\{s_n\}$ is given by (4.4a) with initial values $s_0 = 0$ and $s_1 = B(0)\eta$. Now we show by induction that

(4.26) $$s_{n+1}-s_n \leq t_{n+1}-t_n \quad \text{and} \quad s_{n+1} \leq t_{n+1} \,,$$

where the sequence $\{t_n\}$ is determined by (4.10) or (4.11).

In fact, we have

$$s_{n+1}-s_n = B(s_n)Q((s_n-s_{n-1})/B(s_{n-1}),s_{n-1}) \leq$$

$$\leq B(t_n)Q((t_n-t_{n-1})/B(0),t_{n-1}) \leq$$

$$\leq [1/v(t_n)][u(t_n) - u(t_{n-1}) + v(t_{n-1})(t_n-t_{n-1})] = t_{n+1}-t_n \,,$$

by virtue of (4.22) with $s = t_n$ and $t = t_{n-1}$. It follows from (4.23), that $s_1-s_0 < t_1-t_0$. By Lemma 4.1, the sequence $\{t_n\}$ converges to t^* and all assertions follow from (4.25), (4.26) and Theorem 4.1.

<u>Corollary 4.1</u>. Under the hypotheses of Theorem 4.2, assume instead of (4.22) that

(4.27) $$B(s)Q((s-t)/B(0),t) \leq [1/(1-p_4 s)][p_1(s-t)+p_2+p_3 t](s-t)$$

for $0 \leq t < s \leq t^*$, $n = 1,2,\cdots$, where p_i $(i = 1,\cdots,4)$ satisfy the hypotheses of Lemma 4.3. Then all assertions of Theorem 4.2 hold true provided the sequence $\{t_n\}$ and t^* are defined as in Lemma 4.3 (with $u(t) = \bar{p}_1 t^2 - (1-p_2)t + \eta$ and $v(t) = 1 - p_4 t)$.

Proof. Put $\bar{p}_1 = \max[p_1, (p_3+p_4)/2]$. Then $p_1 \leq \bar{p}_1$, $p_3 \leq 2\bar{p}_1 - p_4$ and (4.27) implies

$$(4.28)\quad B(s)Q((s-t)/B(0),t) \leq [1/(1-p_4 s)][\bar{p}_1(s-t)+p_2+(2\bar{p}_1-p_4)t](s-t) =$$

$$= [1/v(s)][u(s) - u(t) + v(t)](s-t)$$

for $0 \leq t < s \leq t^*$. Hence, it follows that the hypotheses of Theorem 4.2 are satisfied with Lemma 4.3 replacing Lemma 4.1.

Corollary 4.2. Under the hypotheses of Corollary 4.1, let

$$(4.29)\qquad p_2 = p_3 = 0 \quad \text{and} \quad p_4 = 2p_1 = a > 0 .$$

If, in addition, $h = a\eta \leq 1/2$, then all assertions of Theorem 4.2 hold true and the error estimate is as follows,

$$\|x-x_n\| \leq t^* - t_n \leq (a2^n)^{-1}(2h)^{2^n} , \quad n = 0,1,\cdots$$

where $t^* = [1 - (1-2h)^{1/2}]/a$.

Proof. Since $\bar{p}_1 = p_1$ by virtue of (4.29), conditions (4.27) and (4.28) coincide yielding (4.22). Hence, it follows that the hypotheses of Theorem 4.2 are satisfied and the error estimate results from Lemma 4.4.

Remark 4.1. Assume that

$$Q(w,t) = (Kw + d_0 + d_1 t)w$$

and

$$B(t) = B/(1-Ct) .$$

Then,

$$B(s)Q((s-t)/B(0),t) = [1/(1-Ct)][(K/B)(s-t) + d_0 + d_1 t](s-t) .$$

Hence, it follows that condition (4.27) of Corollary 4.1 is satisfied with $p_1 = K/B$, $p_2 = d_0$, $p_3 = d_1$ and $p_4 = C$.

In the case of Corollary 4.2, we have $p_2 = p_3 = 0$. The assumptions of Corollary 4.2 are fulfilled if

$$C = 2K/B \quad \text{and} \quad K\eta/B \leq 1/4 .$$

5. More general contractor inequalities

While the boundedness condition remains the same as in the preceding section, the contractor inequality will be of different type. Thus let $P:D \subset X \to Y$ be a nonlinear operator and let $x_0 \in D$ be chosen so that $S = S(x_0,r) \subset D$, where the radius r will be defined below. Let $\Gamma:S \to L(Y,X)$ be a contractor for P satisfying the following contractor inequality,

$$(5.1) \qquad \|P(x+\Gamma(x)y) - Px - y\| \leq Q(\|\Gamma(x)y\|,\|x-x_0\|)$$

for $x \in S$, $y \in Y$ whenever $x+\Gamma(x)y \in D$. It is assumed that $Q(w,t) \geq 0$ for $w,t \geq 0$ is continuous and nondecreasing in each variable. In order to investigate the iterative procedure (2.4), we introduce the following numerical iterative sequence,

$$(5.2) \qquad t_{n+1} = t_n + B(t_n)Q(t_n-t_{n-1},t_{n-1}) , \quad n = 1,2,\cdots$$

with initial value $t_0 = 0$ and we assume that $t^* = \lim_{n\to\infty} t_n$.

Theorem 5.1. Suppose that $P:D \subset X \to Y$ is a closed nonlinear operator with contractor Γ satisfying (5.1) and (4.2). If $x_0 \in D$ is such that

$$(5.3) \qquad \|\Gamma(x_0)Px_0\| \leq t_1 , \quad r = t^* \quad \text{and} \quad Q(0,t^*) = 0 ,$$

then all x_n lie in S and the sequence $\{x_n\}$ defined by (2.4) converges to a solution x of $Px = 0$ and the error estimate holds,

$$\|x-x_n\| \leq t^*-t_n , \quad n = 0,1,\cdots .$$

Proof. Using (5.1) with $x = x_{n-1}$ and $y = -Px_{n-1}$, we show by induction that

$$\|Px_n\| \leq Q(\|\Gamma(x_{n-1})Px_{n-1}\|, \|x_{n-1}-x_0\|) =$$

$$= Q(\|x_n - x_{n-1}\|, \|x_{n-1}-x_0\|) \leq Q(t_n - t_{n-1}, t_{n-1}) ,$$

$$\|x_n - x_0\| \leq t_n \leq t^* , \tag{5.4}$$

$$\|x_{n+1}-x_n\| \leq B(t_n)Q(t_n-t_{n-1},t_{n-1}) = t_{n+1}-t_n \quad \text{for } n = 1,2,\cdots , \tag{5.5}$$

by virtue of (4.2), (2.4), (5.2) and (5.3). Hence the assertions follow from (5.4) and (5.5) by using a standard argument.

The following is a special case of Theorem 5.1.

Corollary 5.1. Under the hypotheses of Theorem 5.1, assume that there exist two functions $u(t)$ and $v(t)$ satisfying the assumptions of Lemma 4.1 and such that

$$B(s)Q(s-t,t) \leq [1/v(s)][u(s) - u(t) + v(t)(s-t)] \tag{5.6}$$

for $0 \leq t < s \leq t^*$, where t^* is the smallest positive root of $u(t) = 0$. If $r = t^*$, then all assertions of Theorem 5.1 hold true provided the sequence $\{t_n\}$ is defined as in Lemma 4.1.

Proof. By virtue of (5.6), with $s = t_n$ and $t = t_{n-1}$, we obtain

$$B(t_n)Q(t_n-t_{n-1},t_{n-1}) \leq [1/v(t_n)][u(t_n) - u(t_{n-1}) + v(t_{n-1})](t_n-t_{n-1}) =$$

$$= t_{n+1}-t_n$$

for $n = 1,2,\cdots$. Since, by Lemma 4.1, the sequence $\{t_n\}$ converges to t^*, the proof follows from Theorem 5.1.

Corollary 5.2. Under the hypotheses of Theorem 5.1, assume that

(5.7) $$B(s)Q(w,t) \leq [1/(1-p_4 s)][p_1 w + (p_2+p_3 t)]w$$

for $0 \leq t < s \leq t^*$ where $w = s-t$ and p_i $(i = 1,\cdots,4)$ satisfy the assumptions of Lemma 4.3. If

$$\|\Gamma(x_0)Px_0\| \leq \eta \quad \text{and} \quad \eta \leq (1-p_2)^2/4\bar{p}_1 ,$$

then all assertions of Theorem 5.1 hold true provided the sequence $\{t_n\}$ is defined as in Lemma 4.3.

Proof. It follows from (5.7) that

$$B(s)Q(s-t,t) \leq [1/(1-p_4 s)][\bar{p}_1(s-t) + p_2 + (2\bar{p}_1-p_4)t](s-t) .$$

Hence,

$$B(t_n)Q(t_n-t_{n-1},t_{n-1}) \leq$$

$$\leq [1/(1-p_4 t_n)][\bar{p}_1(t_n-t_{n-1}) + p_2 + (2\bar{p}_1-p_4)t_{n-1}](t_n-t_{n-1}) =$$

$$= t_{n+1}-t_n .$$

Since $\{t_n\}$ with initial values $t_0 = 0$, $t_1 = \eta$ is the same as in Lemma 4.3, it converges to t^* and the proof follows from Theorem 5.1.

Corollary 5.3. Under the hypotheses of Theorem 5.1, put in (5.1),

(5.8) $$Q(w,t) = (Kw+d_0+d_1 t)w$$

with d_0 , $d_1 \geq 0$ and in (4.2),

(5.9) $$B(t) = B/(1-Ct)$$

with $Bd_0 < 1$. Let t^* and $\{t_n\}$ be defined as in Lemma 4.3 with $p_1 = BK$, $p_2 = Bd_0$ and $p_4 = C$. If $0 < \eta \leq (1-p_2)^2/4\bar{p}_1$ where $\bar{p}_1 = \max[BK,(Bd_1+C)/2]$. Then all assertions of Theorem 5.1 hold true.

Proof. It follows from Lemma 4.3 that the sequence $\{t_n\}$ which is defined as in the proof of Corollary 5.2, is convergent and $t^* = \lim_{n\to\infty} t_n$. Therefore, the proof follows from Theorem 5.1.

Lemma 5.1. Suppose that $P:D \subset X \to Y$ is differentiable in the sense of Fréchet on $S = S(x_0,r) \subset D$ and

$$\|P'(x) - P'(\bar{x})\| \leq K\|x-\bar{x}\| , \quad x,\bar{x} \in S .$$

Let $A(x):X \to Y$ be a bounded linear operator associated with $x \in D$ and let $x_0 \in D$ be such that with $d_0,d_1 \geq 0$,

$$\|A(x) - A(x_0)\| \leq M\|x-x_0\| ,$$

$$\|P'(x) - A(x)\| \leq d_0 + d_1\|x-x_0\| ,$$

for all $x \in S$. Assume, moreover, that $A(x_0)$ is nonsingular and $\|A(x_0)^{-1}\| \leq B$. Then $\Gamma(x) = A(x)^{-1}$ is a contractor for P with $Q(w,t)$ and $B(t)$ given by (5.8) and (5.9), respectively, with $K/2$ replacing K and $C = BM$ provided $rBM \leq 1$.

Proof. We have, for $x \in S$,

$$\|A(x) - A(x_0)\| \leq M\|x-x_0\| < Mr \leq 1/B .$$

Hence, $A(x)$ is nonsingular, and

$$\|\Gamma(x)\| \leq B/(1-BM\|x-x_0\|) \quad \text{for} \quad x \in S ,$$

where $\Gamma(x) = A^{-1}(x)$. Therefore, we have

$$\|P(x+\Gamma(x)y) - Px - y\| \leq \|P(x+\Gamma(x)y) - Px - P'(x)\Gamma(x)y\| +$$

$$+ \|P'(x)\Gamma(x)y - A(x)\Gamma(x)y\| \leq (1/2)K\|\Gamma(x)y\|^2 + (d_0+d_1\|x-x_0\|)\|\Gamma(x)y\|$$

$$\text{for} \quad x \in S ,$$

whenever $x+\Gamma(x)y \in S$. Thus, $\Gamma(x):Y \to X$ is a contractor for P and satisfies the requirements for $Q(w,t)$ and $B(t)$.

As a special case of Corollary 5.3 (except for uniqueness), we obtain the following theorem which is due to Rheinboldt [1] (see also Ortega and Rheinboldt [2], 12.6.4).

Theorem 5.2. Suppose that $P:D \subset X \to Y$ is differentiable in the sense of Fréchet on the convex set $D_0 \subset D$ and

$$\|P'(x) - P'(\bar{x})\| \le K\|x-\bar{x}\| , \quad x,\bar{x} \in D_0 .$$

Let $A(x):X \to Y$ be a bounded linear operator associated with $x \in D_0$ and let $x_0 \in D_0$ be such that with $d_0, d_1 \ge 0$,

$$\|A(x) - A(x_0)\| \le M\|x-x_0\| ,$$

$$\|P'(x) - A(x)\| \le d_0+d_1\|x-x_0\| ,$$

for all $x \in D_0$. Assume, moreover, that $A(x_0)$ is nonsingular and

$$\|A(x_0)^{-1}Px_0\| \le \eta , \quad \|A(x_0)^{-1}\| \le B$$

and that $Bd_0 < 1$. Let t^* and $\{t_n\}$ be defined as in Lemma 4.3 with $p_1 = BK/2$, $p_2 = Bd_0$, $p_3 = Bd_1$ and $p_4 = C = BM$. If $\bar{S}(x_0,t^*) \subset D_0$ and $0 < \eta \le (1-Bd_0)^2/4\bar{p}1$ where $\bar{p}_1 = \max[BK/2, B(d_1+M)/2]$, then the sequence $\{x_n\}$ defined by

$$x_{n+1} = x_n - A(x_n)^{-1}Px_n , \quad n = 0,1,\cdots$$

remains in $S(x_0,t^*)$ and converges to a solution x of $Px = 0$ and the error estimate holds,

$$\|x-x_n\| \le t^*-t_n , \quad n = 0,1,\cdots .$$

Proof. Since $t^*BM \le BM/2\bar{p}_1 \le 1$ and, by virtue of Lemma 5.1, $\Gamma(x) = A(x)^{-1}$, $x \in S(x_0,t^*)$, is a contractor for P with $Q(w,t)$ and $B(t)$ defined as in Corollary 5.3, where $K/2$ replaces K and $C = BM$. Thus, the hypotheses of Corollary 5.3 are satisfied.

Corollary 5.4. Set in (5.8) of Corollary 5.3, $d_0 = d_1 = 0$. Let t^* and $\{t_n\}$ be defined as in Lemma 4.4 with $h = \eta a$ and $a = 2\bar{p}_1$ where $\bar{p}_1 = \max[BK, C/2]$. If $0 < \eta \le 1/4\bar{p}_1$, then all assertions of Theorem 5.1 hold true and the error estimate is

(5.10) $$\|x-x_n\| \leq t^* - t_n \leq (a2^{n+1})^{-1}(2h)^{2^{n+1}}$$

Proof. Using the sequence $\{t_n\}$ defined in the proof of Corollary 5.2, we obtain

$$\|x_{n+1}-x_n\| \leq t_{n+1}-t_n$$

and

$$\|x_n-x_0\| \leq t_n \ , \quad n = 0,1,\cdots ,$$

by virtue of (5.4) and (5.5). The sequence $\{t_n\}$ is the same as that defined in Lemmas (4.3) and (4.4). By virtue of Lemma (4.4), $\{t_n\}$ is convergent yielding the error estimate (5.10).

As a particular case of Corollary 5.4, we obtain the following theorem which is due to Kantorovič [1]. For references concerning this theorem, see Ortega and Rheinboldt [2]. In the literature, this theorem is called the Newton-Kantorovič Theorem.

Theorem 5.3. Assume that $P:D \subset X \to Y$ is differentiable in the sense of Fréchet in $S = S(x_0,r) \subset D$ and that $\|P'(x) - P'(\bar{x})\| \leq K\|x-\bar{x}\|$ for all $x,\bar{x} \in S$. Suppose that x_0 is such that $\|P'(x_0)^{-1}\| \leq B$ and $\|P'(x_0)^{-1}Px_0\| \leq \eta$, where

$$h = BK\eta \leq 1/2 \quad \text{and} \quad r = t^* = [1-(1-2h)^{1/2}]/BK .$$

Then the iterates

$$x_{n+1} = x_n - P'(x_0)^{-1}Px_n \ , \quad n = 0,1,\cdots$$

remain in S and converge to a solution x of $Px = 0$. Moreover, the error estimate (5.10) holds.

Proof. We have

$$\|P'(x) - P'(x_0)\| \leq K\|x-x_0\| < B^{-1}$$

$$\text{for } x \in S(x_0,r)$$

with $r = (BK)^{-1}$. A standard argument shows that $P'(x)$ is nonsingular in S and $\|P'(x)^{-1}\| \leq B(1 - BK\|x-x_0\|)^{-1}$ for $x \in S$. We also have

$$\|P(x+\Gamma(x)y) - Px - y\| = \|P(x+\Gamma(x)y) - Px - P'(x)\Gamma(x)y\| \leq K\|\Gamma(x)y\|^2/2 ,$$

whenever $x+\Gamma(x)y \in S$, where $\Gamma(x) = P'(x)^{-1}$. Thus, $\Gamma(x)$ is a contractor satisfying inequalities (5.1) with $Q(w,t) = Kw^2/2$ and (5.9) with $C = BK$. It is easily seen that the hypotheses of Corollary 5.4 are satisfied with K replaced by $K/2$ in (5.8).

Notice that under the same hypotheses, Kantorovič [1] has also proved the uniqueness of the solution in the open sphere $S(x_0, r)$ with $r = [1 - (1-2h)^{1/2}]/BK$ if $h < 1/2$ and in the closed sphere $\bar{S}(x_0, (BK)^{-1})$ if $h = 1/2$.

CHAPTER 3

SERIES OF ITERATES AND A CLASS OF MAJORANT FUNCTIONS

Introduction. The subject matter of this chapter, for the most part (sections 1-4), pertains to the classical theory of infinite series established by A. L. Cauchy in the 19th century and further developed by other mathematicians. Also, the technique which is used here is not a product of the modern mathematics and, as a matter of fact, was available even before the principles of the theory of infinite series were established. If some ideas here appear to be new, it can only be surprising.

In the first two sections, series of iterates of a positive function are discussed. Series of iterates constitute a very comprehensive class which contains: geometric series, Dirichlet series, Bertrand series and more general series which appear in the Cauchy integral test. An Ermakoff type test is also proposed for series of iterates.

An equivalent of Cauchy's integral test is discussed in sections 3 and 4. An application to power series is given there. A fixed point theorem which generalizes the contraction principle for complete metric spaces is presented in section 4. It is based upon the equivalent of Cauchy's integral test. Applications of series of iterates to generalized contractions, local and global existence theorems are given in sections 5, 6 and 7.

1. Series of iterates of a positive function

1.1. Given a real valued function Q with the property

a) $$0 < Q(s) < s \quad \text{for} \quad 0 < s \leq s_1 \, ,$$

then the series $\sum_{n=1}^{\infty} s_n$, where

$$s_{n+1} = Q(s_n) ,$$

is called a series of iterates (generated by Q) . A class of convergent series of iterates is defined by Altman [6], [7], [12] and it is shown there that the series $S = \sum_{n=1}^{\infty} s_n$ is convergent and

$$\sum_{i=n}^{\infty} s_i \leq \int_0^{s_n} g(s)$$

if Q also satisfies conditions b) and c), in addition to a),

b) The function $g(s) = s/(s-Q(s))$ is nonincreasing ;

c) $$\int_0^{s_1} g(s)ds < \infty .$$

If conditions a) and b) are satisfied, then condition c) is also necessary for the convergence of the sereis S provided the following condition is satisfied,

d) There exist positive numbers a and d such that

$$[Q(u) - Q(v)]/(u-v) \geq d \quad \text{for all} \quad 0 < v < u < a .$$

In this case the existence of the integrals $\int_\alpha^{s_1} g(s)ds$ is obviously required for $0 < \alpha < s_1$. Condition d) can be replaced by the following one,

d') The function $Q(s)$ is differentiable in an open interval $(0,a)$ and the derivative $Q'(s)$ has a limit as $s \to 0+$.

In fact, since the function $Q(s)/s$ is nonincreasing, by b), and, consequently, by a), it has a positive limit ≤ 1 . Hence, by L'Hospital's rule, the derivative $Q'(s)$ has a positive limit as $s \to 0+$. This implies that condition d) is satisfied.

1.2. Let $y = f(s)$ be a nondecreasing function satisfying the condition

$$0 < f(s) < 1 \quad \text{for} \quad 0 < s \le s_1 \ .$$

Put

$$Q(s) = s(1-f(s)) \quad \text{for} \quad 0 < s \le s_1 \quad \text{and} \quad s_{n+1} = Q(s_n) \ , \quad n = 1,2,\cdots \ .$$

If

$$\int_0^{s_1} [f(s)]^{-1} ds < \infty \ , \quad \text{then} \quad \sum_{n=1}^{\infty} s_n < \infty$$

and

$$\sum_{i=n}^{\infty} s_i \le \int_0^{s_n} [f(s)]^{-1} ds \ .$$

In fact, it is easily seen that conditions a), b) and c) are satisfied with $g(s) = [f(s)]^{-1}$.

As a particular case we obtain the following convergent series of iterates by putting

$$f(s) = s^{\alpha}/(1+\alpha) \quad \text{for} \quad 0 < s \le s_1 < 1$$

and $0 < \alpha < 1$, i.e., $s_{n+1} = Q(s_n)$, where $Q(s) = s(1-s^{\alpha}/(1+\alpha))$ (see Altman [7]).

1.3. Let $y = f(s)$ be a nondecreasing function satisfying the condition $0 < f(s) < 1$ for $0 < s \le s_1$. Assume that the derivative $f'(s)$ exists in some open interval $(0,a)$ and it has a limit as $s \to 0+$. Then the series of iterates $s_{n+1} = Q(s_n)$, where $Q(s) = s(1-f(s))$ is convergent if and only if $\int_0^{s_1} [f(s)]^{-1} ds < \infty$ provided the integrals $\int_{\beta}^{s_1} [f(s)]^{-1} ds$ exist for $0 < \beta < s_1$.

In fact, it is easily seen that conditions a), b) and d') are satisfied and, therefore the assertion follows.

1.4. Suppose that Q is a function satisfying condition a). Suppose in addition, that Q is differentiable and the derivative $Q'(s)$ has a limit as $s \to 0+$ and satisfies condition

(e) $$Q'(s) < Q(s)/s \quad \text{for} \quad 0 < s \le s_1 .$$

Then the series of iterates $s_{n+1} = Q(s_n)$, $n = 1,2,\cdots$ is convergent if and only if condition $\int_0^{s_1} g(s)\,ds < \infty$ holds provided the integrals $\int_\beta^{s_1} g(s)\,ds$ exist for $0 < \beta < s_1$. In fact, condition d') is satisfied and so is condition b), since the derivative of $g(s) = s/(s-Q(s))$ exists and is negative, by virtue of e). It is easy to see that the function $Q(s) = s[1-s^\alpha/(1+\alpha)]$ for $0 < s \le s_1 < 1$ satisfies the hypotheses mentioned above.

1.5. The Dirichlet series $S = \sum_{n=1}^{\infty} 1/n^\alpha$, $1 < \alpha$, is a series of iterates generated by $Q(s) = (s^{-1/\alpha}+1)^{-\alpha} < s$ (see Altman [7]). The function

$$g(s) = 1/[1-Q(s)/s] = 1/[1-1/(1+s^{1/\alpha})]$$

is decreasing. The integral in c) is

$$\int_0^{s_1} g(s)\,ds = \int_0^{s_1} [1-1/(1+s^{1/\alpha})]^{-1} ds = \alpha\int_0^{\infty} \{x^{1+\alpha}[1-(x/(x+1))^\alpha]\}^{-1} dx .$$

A general class of series of iterates can be defined by $S = \sum_{n=1}^{\infty} f(n)$, where f is a positive decreasing continuous function such that $f(x) \to 0$ as $x \to \infty$. In this case we have (see Altman [7])

$$Q(s) = f(f^{-1}(s)+1) , \quad s = f(x) , \quad s_n = f(n) , \quad g(s) = s/[s-f(f^{-1}(s)+1)] .$$

If the function with values $f(x)[f(x) - f(x+1)]^{-1}$ is not decreasing, then g is not increasing and condition b) holds true. If, in addition, f is differentiable, then condition c) can be written in the form

$$\int_0^{s_1} g(s)\,ds = \int_0^{s_1} [s-f(f^{-1}(s)+1)]^{-1} ds = \int_1^{\infty} f(x)\,|f'(x)|\,[f(x) - f(x+1)]^{-1} dx < \infty .$$

Condition d) for Q can be written in the form

$$[f(x+1)-f(y+1)]/[f(x)-f(y)] \geq d > 0$$

for sufficiently large $y > x$. This condition can be replaced by the following one,

> $f'(x+1)/f'(x)$ has a limit as $x \to \infty$, provided the function f is differentiable and $f'(x) \neq 0$.

The Bertrand series (see Smail [1])

$$\sum_n [n\ell_1 n\ell_2 n \cdots \ell_{r-1} n \cdot (\ell_r n)^p]^{-1} ,$$

where $\ell_0 x = x$, $\ell_1 x = \log x, \cdots, \ell_r x = \log \ell_{r-1} x$ provide an important class of convergent series of iterates if $p > 1$. This is a subclass of the series of the form $\sum_n f(n)$ mentioned above.

1.6. Let Q be a function which satisfies condition a) and put

$$Q_1(x) = Q(x) , \quad Q_2(x) = Q(Q(x)), \cdots ,$$
$$Q_{k+1}x = Q(Q_k(x)) , \quad \text{for } k = 1,2,\cdots ,$$

and let $\{b_k\}$ be an increasing sequence of positive integers such that $b_{k+1} - b_k \leq M(b_k - b_{k-1})$. Then the series $\sum_{n=1}^{\infty} s_n$, where $s_{n+1} = Q(s_n)$, and the series $\sum_{k=1}^{\infty} (b_{k+1}-b_k) Q_{b_k}(s_1)$ are either both convergent or both divergent. This fact results immediately from Cauchy's condensation test (see Knopp [1]), since we have by definition that $s_n = Q_n(s_1)$.

2. An Ermakoff type test for series of iterates

2.1. The argument used by Ermakoff (see Knopp [1], p. 296) can be applied to series of iterates in order to obtain a similar test. For this purpose we investigate the convergence of the improper (in general) integral in c). Let Q be continuous and let a) be satisfied. Then we obtain

$$\int_0^1 g(s)\,ds = \int_0^\infty e^{-x} g(e^{-x})\,ds = \int_0^\infty f(x)\,dx = I ,$$

where $f(x) = e^{-x}g(e^{-x})$, $s = e^{-x}$. Put $E(x) = e^{x}f(e^{x})/f(x)$. It follows from the argument of Ermakoff (see Knopp [1], p. 297) that the integral I is convergent if there exists a positive number $\theta < 1$ such that $E(x) \leq \theta$ for all sufficiently large x , and I is divergent if $E(x) \geq 1$ for all sufficiently large x . In terms of g and $s = e^{-x}$, we obtain

$$E(x) = e^{-1/s}g(e^{-1/s})/s^2 g(s) = \mathcal{E}(s)$$

or

$$\mathcal{E}(s) = e^{-2/s}(s-Q(s))/s^3(e^{-1/s}-Q(e^{-1/s})) .$$

Thus, the integral I in c) is convergent if $\mathcal{E}(s) \leq \theta < 1$ for all sufficiently small positive s and I is divergent if $\mathcal{E}(s) \geq 1$ for all sufficiently small positive s .

Let us observe that if $g(e^{-1/s})/g(s)$ or $(s-Q(s))/(e^{-1/s}-Q(e^{-1/s}))$ is bounded, then $\mathcal{E}(s) \leq \theta < 1$ for all sufficiently small positive s , since $e^{-1/s}/s^2 \to 0$ and $e^{-2/s}/s^3 \to 0$ as $s \to 0+$.

2.2. The following Ermakoff type test for series of iterates can now be easily obtained.

Let Q be a continuous function satisfying conditions a), b) and d) or d'). Then the series of iterates generated by Q is convergent if there exists a positive number $\theta < 1$ such that $\mathcal{E}(s) \leq \theta$ for all sufficiently small positive s , and this series is divergent if $\mathcal{E}(s) \geq 1$ for all sufficiently small positive s .

The proof follows immediately from the argument of 1.1. This test can also be applied to some special cases considered above.

Remark 2.1. The function e^{x} in Ermakoff's test can of course be replaced by other functions (see Knopp [1], p. 298). If $\varphi(x)$ is any monotone increasing positive function, everywhere differentiable, for which $\varphi(x) > x$ always, then we can replace $E(x) = e^{x}f(e^{x})/f(x)$ by $E(x) = \varphi'(x)f(\varphi(x))/f(x)$.

Let $\varphi(x_0) = 1$ and put $s = 1/\varphi(x)$. Then we obtain

$$\int_0^1 g(s)\,ds = \int_{x_0}^{\infty} \varphi'(x)[\varphi(x)]^{-2} g(1/\varphi(x))\,dx .$$

Hence, $E(x)$ for the integral function is

$$E(x) = \varphi'(\varphi(x))[\varphi(\varphi(x))]^{-2} g(1/\varphi(\varphi(x)))/[\varphi(x)]^2 g(1/\varphi(x)) .$$

In terms of s we obtain

$$\mathscr{E}(s) = \varphi'(1/s) g(1/\varphi(1/s))/s^2[\varphi(1/s)]^2 g(s) .$$

2.2. Consider the power series $\sum_{n=0}^{\infty} s_n x^n$, where $s_{n+1} = Q(s_n)$ for $n = 0,1,\cdots$. Suppose that the function Q satisfies conditions a) and b) of section 1.2. Then the radius of convergence of the power series is 1 or the series $\sum_{n=0}^{\infty}$ can be majorized by a geometric series.

<u>Proof</u>. It follows from a) and b) that the sequence of $s_{n+1}/s_n = Q(s_n)/s_n < 1$ is nondecreasing, and, therefore, it has a limit $\mu \leq 1$. If $\mu < 1$, then $s_n \leq s_0\mu^n$ for all n. If $\mu = 1$, then, by virtue of a theorem of Cauchy $s_n^{1/n} \to \mu = 1$ as $n \to \infty$. Hence, the radius of convergence equals $\frac{1}{\mu} = 1$.

2.3. An application of Abel's limit theorem (see Knopp [1]) can be given as follows.

Consider the power series $f(x) = \sum_{n=0}^{\infty} s_n x^n$, where $s_{n+1} = Q(s_n)$ with Q satisfying conditions a), b) of section 1.1. Suppose that

(i) $\lim_{s \to 0+} Q'(s) = 1$.

Then $\lim_{x \to 1-0} f(x)$ exists and is equal to $\sum_{n=0}^{\infty} s_n < \infty$ if and only if $\int_0^{s_0} g(s)\,ds < \infty$, and $\lim_{x \to 1-0} f(x) = \infty$ if and only if $I = \infty$, where $g(s) = s/(s-Q(s))$. Condition (i) can be replaced by condition d) and $\sup[Q(s)/s] = 1$.

<u>Proof</u>. Since 1 is the radius of convergence for the power series $f(x)$, the proof follows from Abel's limit theorem and from the fact

that, in this case, condition c) of section 1.1 is necessary and sufficient for the convergence of I .

Example. $Q(s) = s(1-s^{\alpha}/(1+\alpha))$, $0 < \alpha < 1$, $0 < s \leq s_0 < 1$, satisfies all the hypotheses.

3. Cauchy's integral test

3.1. Let f be a continuous nonincreasing function which is positive for $x > 0$. Then, by Cauchy's integral test, the series $\sum_{n=0}^{\infty} f(n)$ is convergent if and only if $\int_0^{\infty} f(x)\,dx < \infty$.

Now let $B(s)$ be a nondecreasing continuous positive function defined on the interval $(0,1]$ and let $q < 1$ be an arbitrary fixed positive number.

Then the series $\sum_{n=1}^{\infty} B(q^n)$ is convergent if and only if

$$\int_0^1 s^{-1}B(s)\,ds < \infty . \tag{3.1}$$

This fact follows immediately from Cauchy's integral test by putting $f(x) = B(q^x)$. Thus, we obtain by putting $s = q^x$ that

$$\int_0^{\infty} f(x)\,dx = |\log q|^{-1}\int_0^1 s^{-1}B(s)\,ds .$$

For any function $f(x)$ which satisfies the assumptions made above for Cauchy's integral test we can always define a function $B(s)$ by the formula $f(x) = B(q^x)$ for $s = q^x$ where q is arbitrary with $0 < q < 1$. The function B has of course all the properties required above. In this simple way we obtain an equivalent form of Cauchy's integral test.

3.2. Consider the power series $\sum_{n=0}^{\infty} a_n x^n$. Let $B(s)$ be a positive nondecreasing continuous function defined on $[0,1]$ and such that $\int_0^1 s^{-1}B(s)\,ds < \infty$. Suppose that ρ and $q < 1$ are positive numbers such that

$$|a_n| \leq B(q^n)/\rho^n$$

for almost all positive integers n .

Then the power series $\sum_{n=0}^{\infty} a_n x^n$ is uniformly convergent for $|x| \leq \rho$.

This fact follows immediately from the equivalent Cauchy integral test.

It turns out that by using the well-known Cauchy-Hadamard theorem in order to find the radius of convergence for the power series considered in this case, one can only prove the uniform convergence for $|x| \leq \bar{\rho}$, where $0 < \bar{\rho} < \rho$.

3.3. An Ermakoff type test in terms of the function $B(s)$ can easily be established. In other words, we have to find a convergence test for the integral $I = \int_0^1 s^{-1}B(s)ds$.

Replacing $g(s)$ in 2.1 by the function $s^{-1}B(s)$ we obtain the required expression for $\delta(s)$,

$$\delta(s) = e^{-1/s}B(e^{-1/s})/sB(s) , \quad 0 < s \leq 1 .$$

Hence, it follows that the integral I is convergent if there exists a positive number $\theta < 1$ such that

$$\delta(s) \leq \theta < 1 \tag{3.2}$$

for all sufficiently small positive s , and the integral I is divergent if

$$\delta(s) \geq 1 \tag{3.3}$$

for all sufficiently small $s > 0$.

Notice that if $B(e^{-1/s})/B(s)$ is bounded for $0 < s \leq 1$, then $\delta(s) \leq \theta < 1$ for sufficiently small $s > 0$, since $e^{-1/s}/s \to 0$ as $s \to 0+$.

There is a more direct approach by investigating the convergence of the integral $I_1 = \int_0^\infty B(q^x)\,dx$, where $q < 1$ is positive. Then we obtain, by Ermakoff's test, that if there exists a positive $\theta < 1$ such that

$$(3.4) \qquad E_q(x) = e^x B(q^{e^x})/B(q^x) \leq \theta < 1$$

for all sufficiently large x , then the integral I_1 is convergent, and if

$$(3.5) \qquad E_q(x) \geq 1$$

for all sufficiently large x , then the integral I_1 is divergent.

Since the convergence of the integral $I_1 = \int_0^\infty B(q^x)\,dx$ for a particular positive $q < 1$ implies the convergence of the integral $I = \int_0^1 s^{-1}B(s)\,ds$ and the convergence of I yields the convergence of I_1 for arbitrary positive $q < 1$, it follows that the convergence (divergence) of I_1 for a particular positive $q < 1$ implies the convergence (divergence) of I_1 for arbitrary positive $q < 1$.

Suppose now that there exist positive numbers $q < 1$ and $\theta < 1$ such that one of the conditions (3.1), (3.2) or (3.4) is satisfied, and assume that $|a_n| \leq B(q^n)/\rho^n$ for almost all positive integers n . Then the power series $\sum_{n=0}^{\infty} a_n x^n$ is absolutely convergent for $|x| \leq \rho$. On the other hand, if there exist positive numbers ρ and $q < 1$ such that one of the conditions (3.3), (3.5) or $I = \infty$ is satisfied and $|a_n| \geq B(q^n)/\rho^n$ for almost all positive integers n , then the power series $\sum_{n=0}^{\infty} a_n x^n$ is absolutely divergent for $|x| \geq \rho$.

Notice that the latter case does not provide information needed to compute the radius of convergence on the basis of the Cauchy-Hadamard theorem.

3.4. Consider now the Taylor series $\sum_{n=0}^{\infty} (1/n!) f^{(n)}(a)(x-a)^n$ for the function f . Let $B(s)$ be a continuous nondecreasing function which is positive for $0 < s \leq 1$ and satisfies condition (3.1). Suppose that there exist positive numbers ρ and $q < 1$ such that $(1/n!)|f^{(n)}(a)| \le B(q^n)/\rho^n$ for almost all positive integers n. Then the Taylor series for f is absolutely uniformly convergent if $|x-a| \le \rho$.

Of course, condition (3.1) can be replaced by condition (3.2) or (3.4).

Suppose now that the integral $I = \int_0^1 s^{-1}B(s)ds = \infty$ and $\rho > 0$, $0 < q < 1$ are such that $(1/n!)|f^{(n)}(a)| \ge B(q^n)/\rho^n$ for almost all positive integers n. Then the Taylor series for f is absolutely divergent if $|x-a| \ge \rho$.

Of course, the assumption that the integral $I = \infty$ can be replaced by condition (3.3) or (3.5).

4. A fixed point theorem

4.1. As an application of the equivalent of Cauchy's integral test we obtain the following fixed point theorem.

Let $F:X \to X$ be a continuous mapping of the complete metric space X into itself. Let $B(s) > 0$ for $0 < s \le a$ be a continuous non-decreasing function such that $\int_0^a s^{-1}B(s)ds < \infty$. Suppose that there exist an element $x \in X$ and a positive number $q < 1$ such that $d(Fx,x) \le a$ and

$$(4.1) \qquad d(F^{n+1}x,F^nx) \le B(q^n d(Fx,x)) \quad \text{for } n = 1,2,\cdots,$$

where $d(\cdot,\cdot)$ is the distance in X. Put $x_{n+1} = Fx_n$ for $n = 1,2,\cdots$. Then the sequence $\{x_n\}$ converges to a fixed point $x^* = Fx^*$.

Proof. We have for $m > n$, $d(x_{m+1},x_{n+1}) = d(F^mx,F^nx) \le$ $\le \sum_{i=n}^{m-1} d(F^{i+1}x,F^ix) \le \sum_{i=n}^{m-1} B(q^id(Fx,x))$. Since the series $\sum_{i=1}^{\infty} B(q^id(Fx,x))$ is convergent (see section 3.1), it follows that the sequence of elements $x_{n+1} = Fx_n$ is a Cauchy sequence and, therefore, it has a limit x^*. The continuity of F implies that $x^* = Fx^*$.

Remark 4.1. Suppose that F is a contraction mapping, that is, there exists a positive constant $q < 1$ such that

$$d(Fx,Fy) \leq qd(x,y) \quad \text{for all} \quad x,y \in X .$$

Then, of course, F satisfies condition (4.1) if we put $B(s) = s$ and x is an arbitrary element of X.

Notice that this theorem actually yields an application of the Cauchy's integral to the general theory of fixed points. The Cauchy integral test appears in its equivalent form discussed in section 3.1.

As to the rate of convergence of the sequence of the successive approximations $x_{n+1} = Fx_n$, the following observation is immediate. If F is a contraction mapping, then the convergence is linear, that is the rate of convergence is the same as that for some geometric series. However, in general the rate depends essentially on the function B.

5. Generalized contractions

Let $F:X \to X$ be a mapping such that $d(Fx,Fy) \leq Q(d(x,y))$ for all x,y (or with $y = Fx$) of the complete metric space X with distance d. Put $x_{n+1} = Fx_n$, $n = 0,1,\cdots$ and $s_{n+1} = Q(s_n)$ with $s_1 = d(x_1,x_0)$, where Q is a nondecreasing function satisfying conditions a), b) and c). Then the sequence $\{x_n\}$ converges to a unique fixed point $x = Fx$.

Proof. We prove by induction that $d(x_n,x_{n+1}) = d(Fx_{n-1},Fx_n) \leq \leq Q(d(x_{n-1},x_n)) \leq Q(s_n) = s_{n+1}$. Hence, we have

$$(5.1) \qquad d(x_{n-1},x_{m-1}) \leq \sum_{i=n}^{m-1} d(x_i,x_{i+1}) \leq \sum_{i=n}^{m} s_{i+1} \leq \int_{s_m}^{s_n} g(s)ds ,$$

by 1.1, and $\{x_n\}$ converges to some $x \in X$. Since $d(Fy,Fz) \leq \leq Q(d(y,z)) < d(y,z)$, the mapping F is continuous and $Fx = x$. If $x \neq y = Fy$, then $0 < d(x,y) = d(Fx,Fy) \leq Q(d(x,y)) < d(x,y)$, yielding a contradiction. Letting $m \to \infty$ in (5.1) we obtain the error estimate $d(x_{n-1},x) \leq \int_0^{s_n} g(s)ds$.

For generalized contractions, see also Boyd and Wong [1].

6. Natural majorant functions

Let $P: D \subset X \to Y$ be a non-linear operator with domain D containing a ball $S = S(x_0, r)$ with radius r and center x_0, X and Y being Banach spaces. Denote by $L(Y \to X)$ the space of all linear bounded operators from Y into X and let $\Gamma: D \to L(Y \to X)$ be a mapping, that is, for fixed $x \in D$, $\Gamma(x): Y \to X$ is a bounded linear operator from Y into X. Put

$$(6.1) \quad Q(s) = \max\{\|P(x+\Gamma(x)y) - Px - y\| \mid x \in S,\ y \in Y,\ x+\Gamma(x)y \in D,\ \|y\| \leq s\},$$

assuming that $Q(s)$ is finite for $0 < s \leq \eta$, where $\eta > 0$ is a certain number to be defined below. It follows from (6.1) that $Q(0) = 0$ and $Q(s)$ is non-decreasing. It results also from (6.1) that

$$(6.2) \qquad \|P(x+\Gamma(x)y) - Px - y\| \leq Q(\|y\|)$$

for $x \in S$, $y \in Y$ whenever $x+\Gamma(x)y \in D$. If there exists a function Q satisfying (6.2), then, of course, Γ is a contractor for P with majorant function Q satisfying the contractor inequality (6.2). Thus, a majorant function for Γ, if any, can be defined in a natural way by (6.1).

Consider now the problem of solving the operator equation

$$(6.3) \qquad Px = 0, \quad x \in D.$$

We assume that x_0 is chosen so as to satisfy

$$(6.4) \qquad \|Px_0\| \leq \eta$$

and Γ is bounded; that is

$$(6.5) \qquad \|\Gamma(x)y\| \leq B \quad \text{for} \quad x \in S.$$

In addition, we suppose that

$$(6.6) \qquad Q(s) < s \quad \text{for} \quad s > 0$$

and there exists the integral

$$I(\eta) = \int_0^{\eta} s[s-Q(s)]^{-1}ds < \infty \tag{6.7}$$

provided the function $s/(s-Q(s))$ is non-increasing.

In order to solve equation (6.3), we use the following iterative procedure

$$x_{n+1} = x_n - \Gamma(x_n)Px_n, \quad n = 0,1,\cdots. \tag{6.8}$$

Simultaneously we consider the following numerical iterative procedure

$$s_{n+1} = Q(s_n), \quad s_0 = \eta, \quad n = 0,1,\cdots, \tag{6.9}$$

where Q is the majorant function for Γ defined by (6.1) or (6.2).

<u>Lemma 6.1</u>. Let $Q(s) > 0$ for $s > 0$ with $Q(0) = 0$ be a function satisfying conditions (6.6) and (6.7), where $s/(s-Q(s))$ is non-increasing. Then the series $\sum_{i=0}^{\infty} s_i$ is convergent, where the sequence $\{s_n\}$ is defined by (6.9) and the following estimate holds,

$$\sum_{i=n}^{\infty} s_i \le \sum_{i=n}^{\infty} \int_{s_{i+1}}^{s_i} s[s-Q(s)]^{-1}ds = \int_0^{s_n} s[s-Q(s)]^{-1}ds. \tag{6.10}$$

<u>Proof</u>. We have for $m > n$, by (6.9),

$$\begin{aligned} \sum_{i=n}^{m-1} s_i &= \sum_{i=n}^{m-1} s_i(s_i - s_{i+1})/(s_i - Q(s_i)) \le \sum_{i=n}^{m-1} \int_{s_{i+1}}^{s_i} s[s-Q(s)]^{-1}ds \\ &= \int_{s_m}^{s_n} s[s-Q(s)]^{-1}ds, \end{aligned} \tag{6.11}$$

since the function $s/(s-Q(s))$ is non-increasing by assumption. Inequality (6.10) follows from (6.11), and the convergence of the series $\sum_{i=0}^{\infty} s_i$ results from (6.7), and we have

$$\sum_{i=0}^{\infty} s_i \le I(\eta). \tag{6.12}$$

Lemma 6.1 will be applied to prove the following.

Theorem 6.1. Let $P:D \to Y$ be a closed non-linear operator with domain D containing the ball S. Suppose that Γ is a contractor for P and satisfies condition (6.5) and the contractor inequality (6.2). Furthermore, assume that the majorant function Q satisfies conditions (6.6) and (6.7), where η is given by (6.4). Finally, let

(6.13) $$BI(\eta) = r .$$

Then, all x_n lie in S and the sequence $\{x_n\}$ defined by (6.8) converges to a solution x of equation (6.3) and the error estimate is as follows,

(6.14) $$\|x-x_n\| \le B\int_0^{s_n} s(s-Q(s)]^{-1}ds .$$

Proof. It follows from the contractor inequality (6.2) with $x = x_n$ and $y = -Px_n$ that

(6.15) $$\|Px_{n+1}\| \le Q(\|Px_n\|) .$$

Since Q is non-decreasing we prove by induction, by virtue of (6.15), that

(6.16) $$\|Px_{n+1}\| \le Q(s_n) = s_{n+1} , \quad n = 0,1,\cdots .$$

It follows from (6.8), (6.5) and (6.16) by using induction, that

$$\|x_{n+1}-x_n\| \le Bs_n , \quad n = 0,1,\cdots .$$

Hence it follows, by virtue of Lemma 6.1, and by (6.11), that

(6.17) $$\|x_m-x_n\| \le B\sum_{i=n}^{m-1} s_i \le B\int_{s_m}^{s_n} s[s-Q(s)]^{-1}ds ,$$

and

(6.18) $$\|x_n-x_0\| \le B\sum_{i=0}^{n-1} s_i \le B\int_{s_n}^{\eta} s[s-Q(s)]^{-1}ds < BI(\eta) = r ,$$

by virtue of (6.13). Thus, it results from (6.18) that $x_n \in S$ for

$n = 0,1,\cdots$ and (6.17) shows that the sequence $\{x_n\}$ converges to some element x. On the other hand, by (6.12), the series $\sum_{n=0}^{\infty} s_n$ is convergent and obviously $s_n \to 0$ as $n \to \infty$. Therefore, (6.16) implies that $Px_n \to 0$ as $n \to \infty$. Since P is closed, it follows that $Px = 0$. The error estimate (6.14) follows from (6.17) by letting $m \to \infty$ so that $s_m \to 0$, and the proof is complete.

It is easily seen that the case of a contractor with linear majorant function, investigated in Chapter 1, is a particular one of Theorem 6.1. In other words, we have the following.

Corollary 6.1. Under the corresponding hypotheses of Theorem 6.1, suppose that the majorant function Q is defined by $Q(s) = qs$ with $0 < q < 1$. Then all assertions of Theorem 6.1 hold true and the error estimate (6.14) yields

$$\|x - x_n\| \le B\,\eta q^n/(1-q) \ . \tag{6.19}$$

Proof. Condition (6.6) is obviously fulfilled and $Q(0) = 0$. The function Q is evidently increasing. Since the function $s/(s-Q(s)) = 1/(1-q)$ is constant, $I(\eta) = \eta/(1-q) < \infty$, yielding condition (6.7). By virtue of (6.9), we have

$$s_n = \eta\, q^n, \quad n = 0,1,\cdots .$$

Hence, it follows that in this particular case the error estimate (6.14) coincides with (6.19) and condition (6.13) is replaced by $B\eta/(1-q) = r$.

Example. Let $F: S \to S \subset X$ be a contraction with Lipschitz constant $q < 1$, that is

$$\|Fx - F\bar{x}\| \le q\|x - \bar{x}\| \quad \text{for all } x,\bar{x} \in S .$$

Put $Px = x - Fx$. Then it is easy to verify that $\Gamma(x) \equiv I$ (the identity mapping of X) is a contractor for P with majorant function Q defined by $Q(s) = qs$.

Thus, Theorem 6.1, as well as Corollary 6.1, generalize the well known local Banach contraction principle.

Remark 6.1. Suppose that, in addition to the hypotheses of Theorem 1.1, the contractor $\Gamma(x)$ maps Y onto X. Then the solution of equation (6.3) is unique.

Proof. If x and $\bar{x}$ are two solutions of equation (6.3), then there exists an element $y \in Y$ such that $\bar{x} = x+\Gamma(x)y$, since $\Gamma(x)$ is onto. It follows from the contractor inequality (6.2) that $\|y\| < Q(\|y\|) < \|y\|$, if $\|y\| > 0$, by virtue of (6.6). Hence, we obtain a contradiction which proves that $\|y\| = 0$, that is $\bar{x} = x$.

7. Global existence and convergence theorem

Theorem 6.1 yields a local existence and convergence theorem. However, using the same argument, one can obtain a global existence and convergence theorem. Let $P:D \subset X \to Y$ be a nonlinear operator, D being a vector space, and let $\Gamma:D \to L(Y \to X)$. We assume that $\Gamma(x)(Y) \subset D$ for all $x \in D$. Then we can replace (6.1) by

$$(7.1) \qquad Q(s) = \max\{\|P(x+\Gamma(x)y)-Px-y\| \mid x \in D,\ y \in Y,\ \|y\| \le s\}$$

assuming that $Q(s)$ is finite for arbitrary $s > 0$. Then the contractor inequality (6.2) is replaced by the following one

$$(7.2) \qquad \|P(x+\Gamma(x)y) - Px - y\| \le Q\|y\|)$$

for $x \in D$ and arbitrary $y \in Y$. Condition (6.5) should be replaced by

$$(7.3) \qquad \|\Gamma(x)\| \le B \quad \text{for} \quad x \in D .$$

As in Section 6, we assume that the majorant function Q for Γ satisfies condition (6.6) and that the function $s/(s-Q(s))$ is nonincreasing for $0 \le s < \infty$. Finally, we assume that there exists the integral

$$(7.4) \qquad I(a) = \int_0^a s[s-Q(s)]^{-1}ds < \infty$$

for arbitrary positive a.

Now we can prove the following global existence and convergence

Theorem 7.1. Let $P:D \to Y$ be a closed non-linear operator and let Γ be a bounded contractor for P satisfying the contractor inequality (7.2), condition (7.3) and $\Gamma(x)(Y) \subset D$ for arbitrary $x \in D$. Let the majorant function Q be non-decreasing and satisfy condition (6.6) and let the function $s/(s-Q(s))$ be non-increasing for $0 \leq s < \infty$. If the integral (7.4) exists for arbitrary $a > 0$, then P maps D onto the whole of Y and the sequence $\{x_n\}$ defined by

$$(7.5) \qquad x_{n+1} = x_n - \Gamma(x_n)[Px_n - \ell] \ , \quad n = 1,2,\cdots ,$$

where x_0 is an arbitrary initial approximation, converges to a solution x of $Px - \ell = 0$, where ℓ is an arbitrary element of Y. The error estimate (6.14) holds true, where the sequence $\{s_n\}$ is defined by (6.9) with $\|Px_0 - \ell\| \leq \eta = s_0$.

Proof. Since, for arbitrary $\ell \in Y$, the operators defined by Px and $Px - \ell$ have the same contractor Γ, it is sufficient to show that $Px = 0$ has a solution x and that the sequence $\{x_n\}$ defined by (7.5) with $\ell = 0$ converges to x. Using the same argument as in the proof of Theorem 6.1, we prove (6.16) and (6.17) by induction. By virtue of Lemma 6.1, the series $\sum_{n=0}^{\infty} s_n$ is convergent. Hence, it follows from (6.17) that the sequence $\{x_n\}$ converges to some element x. Then all assertions follow in the same way as in the proof of Theorem 6.1.

Remark 7.1. Under the hypotheses of Theorem 7.1, if, in addition, $\Gamma(x)$ is onto for every $x \in D$, then P is a one-to-one mapping onto the whole of Y.

Proof. The proof is the same as that of Remark 6.1.

Corollary 7.1. If the majorant function Q in Theorem 7.1 is defined by $Q(s) = qs$ with $0 < q < 1$, then all assertions of Theorem 7.1 hold true and the error estimate is given by (6.19) where $\|Px_0 - \ell\| \leq \eta$.

Proof. The proof follows from that of Corollary 6.1.

Example. Let $F:X \to X$ be a contraction with Lipschitz constant $q < 1$. Then $\Gamma(x) \equiv I$ is a contractor for P defined by $Px = x - Fx$ with majorant function $Q(s) = qs$. Thus, both Theorem 7.1 and Corollary 7.1 generalize the well known global Banach contraction principle. The case of contractors with linear majorant functions is investigated in Chapter 1. More facts about contractors with non-linear majorant functions are presented in Chapter 2.

Remark 7.2. The class of majorant functions satisfying the hypotheses of Lemma 6.1 contains all linear majorant functions Q defined by $Q(s) = qs$ with $0 < q < 1$. It is easily seen that majorant functions Q which cannot be majorized by linear ones necessarily possess the following property,

$$(\alpha) \qquad s/(s-Q(s)) \to \infty \quad \text{as} \quad s \to 0 .$$

In fact, since the function $s/(s-Q(s))$ is non-increasing by assumption, it must be bounded by some positive constant M, if condition (α) is not satisfied. But $s/(s-Q(s)) \leq M$ implies $Q(s) \leq qs$ with $q = (M-1)/M < 1$ and $M > 1$ so that Q can be majorized by a linear function with $q < 1$.

Let us observe that the error estimate (6.14) is more accurate that the error estimate (6.19) obtained by replacing if possible the majorant function Q by a linear one with $q = (M-1)/M$.

CHAPTER 4

DIRECTIONAL CONTRACTORS

Introduction. Directional contractors extend the notion of contractors discussed in Chapter 1. In a similar way, we can introduce the concept of a Gâteaux inverse derivative which is a generalization of the inverse of the Gâteaux derivative in a Banach space. We have shown that, by using contractors we can find constructive methods for solving general equations. However, to prove only existence theorems, it is sufficient to define a weaker kind of contractor which is called a directional contractor. But then we obtain a semi-iterative method, i.e., we define iterations with small steps. To handle such iterations we apply a transfinite induction argument used by Gavurin [1] in a particular case. In a very far going generalization this technique will later be considerably developed in conjunction with the concept of contractor directions.

1. Weak inverse derivatives

Let $P:D(P) \subset X \to Y$ be a nonlinear operator from a Banach space X to a Banach space Y, where the domain $D(P)$ is a vector space. Let $\Gamma(x):Y \to X$ be a bounded linear operator such that $\Gamma(x)(Y) \subset D(P)$. Suppose that

$$(1.1) \qquad P(x+t\Gamma(x)y) - Px - ty = o(t) \quad \text{for every} \quad y \in Y ,$$

where $\|o(t)\|/t \to 0$ as $t \to 0$. Then $\Gamma(x)$ is called a Gâteaux inverse derivative of P at $x \in D(P)$. It follows from this definition that $\Gamma(x)$ is one-to-one, and if P has a nonsingular Gâteaux derivative $P'(x)$, then the inverse Gâteaux derivative is uniquely defined and we have $\Gamma(x) = P'(x)^{-1}$.

2. Bounded directional contractors

Let $P:D(P) \subset X \to Y$ be a nonlinear operator from a Banach space X to a Banach space Y, $D(P)$ being a vector space, and let $\Gamma(x):Y \to X$ be a bounded linear operator associated with $x \in D(P)$.

Definition 2.1. Suppose that there exist positive numbers $q = q(P) < 1$ and $\epsilon = \epsilon(x,y) \leq 1$ such that

$$\|P(x+\epsilon\Gamma(x)y) - Px - \epsilon y\| \leq q\epsilon\|y\| \quad \text{for all} \quad y \in Y . \tag{2.1}$$

Then we say that $\Gamma(x)$ is a directional contractor for P at $x \in D(P)$. It follows from this definition that $\Gamma(x)y = 0$ implies $y = 0$, i.e., $\Gamma(x)$ is one-to-one, and an inverse Gâteaux derivative is obviously a directional contractor.

Remark 2.1. It is clear that ϵ in (2.1) can be made arbitrarily small.

Denote by $L(Y \to X)$ the set of all linear continuous mappings of Y into X. Then $\Gamma:D(P) \subset X \to L(Y \to X)$ is called a directional contractor for P. In particular, if there exists a constant B such that

$$\|\Gamma(x)\| \leq B \quad \text{for all} \quad x \in D(P) ,$$

then Γ is called a bounded directional contractor for P.

3. Global existence theorems

In order to apply the transfinite induction method, we make use of the following two lemmas of Gavurin [1].

Lemma 3.1. Let α be an ordinal number of first or second class and let $\{t_\gamma\}_{0\leq\gamma\leq\alpha}$ be a naturally well-ordered sequence of real numbers provided, for ordinal numbers β of second kind (= limit number), we have

$$t_\beta = \lim_{\gamma \nearrow \beta} t_\gamma .$$

Then the following equality holds,

$$t_\alpha = t_0 + \sum_{0\le\gamma<\alpha} (t_{\gamma+1}-t_\gamma) .$$

Lemma 3.2. Let α be an ordinal number of first or second class and let $\{x_\gamma\}_{0\le\gamma\le\alpha}$ be a well-ordered sequence of elements of X provided

$$x_\beta = \lim_{\gamma\nearrow\beta} x_\gamma .$$

Then

$$\|x_\alpha - x_0\| \le \sum_{0\le\gamma<\alpha} \|x_{\gamma+1}-x_\gamma\| .$$

If X is a metric space with distance $d(\cdot,\cdot)$, then we have

$$d(x_\alpha,x_0) \le \sum_{0\le\gamma<\alpha} d(x_{\gamma+1},x_\gamma) .$$

Theorem 3.1. A closed nonlinear operator $P:D(P)\subset X\to Y$ which has a bounded directional contractor Γ is a mapping onto Y.

Proof. Since, for arbitrary fixed $y\in Y$, the operators Px and $Px-y$ have the same bounded directional contractor $\Gamma(x)$, it is sufficient to prove that the equation $Px = 0$ has a solution. To prove this, we construct well-ordered sequences of numbers t_α and elements $x_\alpha \in D(P)$ as follows. Put $t_0 = 0$ and let x_0 be an arbitrary element of $D(P)$. Suppose that t_γ and x_γ have been constructed for all $\gamma<\alpha$, provided, for arbitrary ordinal number $\gamma<\alpha$, inequality (3.1_γ) is satisfied,

$$(3.1_\gamma) \qquad \|Px_\gamma\| \le e^{-(1-q)t_\gamma}\|Px_0\| ;$$

for first kind ordinal numbers $\gamma+1<\alpha$, the following inequalities are satisfied,

$$(3.2_{\gamma+1}) \qquad \|x_{\gamma+1}-x_\gamma\| \le B\|Px_0\|e^{-(1-q)t_\gamma}(t_{\gamma+1}-t_\gamma) ,$$

$$(3.3_{\gamma+1}) \qquad \|Px_{\gamma+1}-Px_\gamma\| \le (1+q)\,\|Px_0\|e^{-(1-q)t_\gamma}(t_{\gamma+1}-t_\gamma) ,$$

$$(3.4_{\gamma+1}) \qquad 0 < \tau_{\gamma+1} = t_{\gamma+1}-t_\gamma < (1-q)^{-1}\ell n(1-q)(1-\bar{q})^{-1} ,$$

where $\bar{q}$ is arbitrary with $q<\bar{q}<1$; and for second kind (limit) ordinal numbers $\gamma < \alpha$, the following relations hold,

$$t_\gamma = \lim_{\beta \nearrow \gamma} t_\beta , \quad x_\gamma = \lim_{\beta \nearrow \gamma} x_\beta , \quad Px_\gamma = \lim_{\beta \nearrow \gamma} Px_\beta . \tag{3.5$_\gamma$}$$

Then it follows from (3.2), (3.4), (3.5), Lemmas 3.1 and 3.2, that for arbitrary $\gamma < \alpha$ and $\lambda < \alpha$ we have

$$\|x_\gamma - x_\lambda\| \le \sum_{\lambda \le \beta < \gamma} \|x_{\beta+1} - x_\beta\| < B\|Px_0\| \sum_{\lambda \le \beta < \gamma} e^{-(1-q)t_\beta}(t_{\beta+1}-t_\beta) =$$

$$= B\|Px_0\| \sum_{\lambda \le \beta < \gamma} e^{(1-q)(t_{\beta+1}-t_\beta)} e^{-(1-q)t_{\beta+1}}(t_{\beta+1}-t_\beta) <$$

$$< (1-q)(1-\bar{q})^{-1}B\|Px_0\| \sum_{\lambda \le \beta < \gamma} e^{-(1-q)t_{\beta+1}}(t_{\beta+1}-t_\beta) <$$

$$< (1-q)(1-\bar{q})^{-1}B\|Px_0\| \sum_{\lambda \le \beta < \gamma} \int_{t_\beta}^{t_{\beta+1}} e^{-(1-q)t}dt = \tag{3.6}$$

$$= (1-q)(1-\bar{q})^{-1}B\|Px_0\| \int_{t_\lambda}^{t_\gamma} e^{-(1-q)t}dt .$$

In the same way, we obtain from (3.3) - (3.5), Lemmas 3.1 and 3.2 that

$$\|Px_\gamma - Px_\lambda\| \le \sum_{\lambda \le \beta < \gamma} \|Px_{\beta+1} - Px_\beta\| \le$$

$$\le (1+q)\|Px_0\| \sum_{\lambda \le \beta < \gamma} e^{-(1-q)t_\beta}(t_{\beta+1}-t_\beta) =$$

$$= (1+q)\|Px_0\| \sum_{\lambda \le \beta < \gamma} e^{(1-q)(t_{\beta+1}-t_\beta)} e^{-(1-q)t_{\beta+1}}(t_{\beta+1}-t_\beta) <$$

$$< (1+q)(1-q)(1-\bar{q})^{-1}\|Px_0\| \sum_{\lambda \le \beta < \gamma} e^{-(1-q)t_{\beta+1}}(t_{\beta+1}-t_\beta) < \tag{3.7}$$

$$< (1-q^2)(1-\bar{q})^{-1}\|Px_0\| \sum_{\lambda \le \beta < \gamma} \int_{t_\beta}^{t_{\beta+1}} e^{-(1-q)t}dt =$$

$$= (1-q^2)(1-\bar{q})^{-1}\|Px_0\| \int_{t_\lambda}^{t_\gamma} e^{-(1-q)t}dt .$$

Suppose that α is a first kind number. If $Px_{\alpha-1} = 0$, then the proof of the theorem is completed.

If $Px_{\alpha-1} \neq 0$, then we put

$$t_\alpha = t_{\alpha-1}+\tau_\alpha, \quad x_\alpha = x_{\alpha-1} - \tau_\alpha\Gamma(x_{\alpha-1})Px_{\alpha-1}, \tag{3.8}$$

where τ_α is chosen so as to satisfy (3.4) and (2.2) with $y = -Px_{\alpha-1}$, i.e.,

$$\|P(x_{\alpha-1}-\tau_\alpha\Gamma(x_{\alpha-1})Px_{\alpha-1}) - Px_{\alpha-1}+\tau_\alpha Px_{\alpha-1}\| \leq q\tau_\alpha\|Px_{\alpha-1}\| . \tag{3.9}$$

Then we obtain, by using (3.9) and the induction assumption $(3.1_{\alpha-1})$,

$$\begin{aligned} \|Px_\alpha\| &\leq (1-\tau_\alpha)\|Px_{\alpha-1}\| + q\tau_\alpha\|Px_{\alpha-1}\| = \\ &= (1-(1-q)\tau_\alpha)\|Px_{\alpha-1}\| < e^{-(1-q)\tau_\alpha}\|Px_{\alpha-1}\| \leq e^{-(1-q)t_\alpha}\|Px_0\| , \end{aligned} \tag{3.10_α}$$

by (3.8). It follows from (3.8) and (3.1_α) that

$$\|x_\alpha-x_{\alpha-1}\| \leq B\tau_\alpha\|Px_{\alpha-1}\| < B(t_\alpha-t_{\alpha-1})e^{-(1-q)t_{\alpha-1}}\|Px_0\| . \tag{3.11_α}$$

By virtue of (3.8), (3.9) and (3.1_α), we obtain

$$\|Px_\alpha-Px_{\alpha-1}\| \leq (1+q)\tau_\alpha\|Px_{\alpha-1}\| \leq (1+q)\|Px_0\|e^{-(1-q)t_{\alpha-1}}(t_\alpha-t_{\alpha-1}). \tag{3.12_α}$$

Thus, conditions (3.1_α), (3.2_α) and (3.3_α) are satisfied for t_α and x_α.

Now, suppose that α is an ordinal number of second kind and put $t_\alpha = \lim_{\gamma\nearrow\alpha} t_\gamma$. Let $\{\gamma_n\}$ be an increasing sequence convergent to α. It follows from (3.6) that

$$\|x_{\gamma_{n+p}}-x_{\gamma_n}\| \to 0 \quad \text{as} \quad n \to \infty .$$

Hence, the sequence $\{x_{\gamma_n}\}$ has a limit x_α and so does $\{x_\gamma\}$. It follows from (3.7) that the sequence $\{Px_{\gamma_n}\}$ has a limit y_α and so does $\{Px_\gamma\}$. Since P is closed, we infer that $x_\alpha \in D(P)$ and

$y_\alpha = Px_\alpha$. If $t_\alpha < +\infty$, then the limit passage in (3.1_{γ_n}) yields (3.1_α). The relationships (3.5_α) are satisfied by the definition of t_α and x_α, since $y_\alpha = Px_\alpha$. The process will terminate if $t_\alpha = +\infty$, where α is of second kind. In this case (3.1_α) yields $Px_\alpha = 0$ and the proof is completed.

The following lemma is obvious.

Lemma 3.3. Let $P:X \to Y$ be a nonlinear operator, X and Y being Banach spaces. Suppose that $\Gamma(x):Y \to X$ is a bounded linear operator and that there exist positive numbers r and $q < 1$ such that

$$\|P(x+\Gamma(x)y) - Px - y\| \leq q\|y\| \quad \text{for all } y \in Y \text{ with } \|y\| \leq r . \tag{3.13}$$

Then $\Gamma(x)$ is a directional contractor for P at $x \in X$.

Lemma 3.4. Let $P:X \to Y$ be a nonlinear operator differentiable in the Fréchet sense with Hölder continuous derivative $P'(x)$, i.e., there exist positive numbers K, $\alpha \leq 1$ such that

$$\|P'(x) - P'(\bar{x})\| \leq K\|x-\bar{x}\|^\alpha \quad \text{for all } x,\bar{x} \in X . \tag{3.14}$$

Moreover, for every $x \in X$, let $A(x):X \to Y$ be a bounded linear nonsingular operator such that

$$\|A(x)^{-1}\| \leq B \quad \text{and} \quad \|P'(x) - A(x)\| \leq C , \quad BC < 1 \tag{3.15}$$

for all $x \in X$ and some constants B, C. Then Γ defined by $\Gamma(x) = A(x)^{-1}$ is a bounded directional contractor for P.

Proof. We have, by (3.14) and (3.15),

$$\begin{aligned} \|P(x+\Gamma(x)y) - Px - y\| &\leq \|P(x+\Gamma(x)y) - Px - P'(x)\Gamma(x)y\| + \\ &+ \|P'(x)\Gamma(x)y - A(x)\Gamma(x)y\| \leq (1+\alpha)^{-1}K\|\Gamma(x)y\|^{1+\alpha} + \\ &+ C\|\Gamma(x)y\| \leq (1+\alpha)^{-1}KB^{1+\alpha}\|y\|^{1+\alpha} + CB\|y\| \leq q\|y\| , \end{aligned}$$

where q is arbitrary with $BC < q < 1$ and $(1+\alpha)^{-1}KB^{1+\alpha}\|y\|^{\alpha} + BC \leq q$ if $\|y\| \leq r$, i.e., $r^{\alpha} = (q-BC)(1+\alpha)/KB^{1+\alpha}$. Hence, by Lemma 3.3, Γ is a bounded directional contractor for P .

Theorem 3.2. Under the hypotheses of Lemma 3.4, P is a mapping onto Y .

Proof. The proof follows immediately from Theorem 3.1 and Lemma 3.4.

Lemma 3.5. Let $P:X \to Y$ be a nonlinear operator and let $T:X \to Y$ be differentiable in the Fréchet sense at each $x \in X$. Moreover, suppose that

$$\|T'(x) - T'(\bar{x})\| \leq K\|x-\bar{x}\|^{\alpha} , \quad 0 < \alpha \leq 1 ,$$

$T'(x)$ being nonsingular with $\|T'(x)^{-1}\| \leq B$ for all $x \in X$, and T being a Lipschitz approximation to P , i.e.,

$$\|(Px-Tx) - (P\bar{x}-T\bar{x})\| \leq C\|x-\bar{x}\| \quad \text{for all} \quad x,\bar{x} \in X ,$$

where C is a constant such that $BC < 1$.

Then Γ defined by $\Gamma(x) = T'(x)^{-1}$ is a bounded directional contractor for P .

Proof. We have

$$\|P(x+\Gamma(x)y) - Px - y\| \leq \|T(x+\Gamma(x)y) - Tx - T'(x)\Gamma(x)y\| +$$

$$+ \|[P(x+\Gamma(x)y) - T(x+\Gamma(x)y)] - [Px-Tx]\| \leq$$

$$\leq (1+\alpha)^{-1}K\|\Gamma(x)y\|^{1+\alpha} + C\|\Gamma(x)y\| \leq (1+\alpha)^{-1}KB^{1+\alpha}\|y\|^{1+\alpha} + CB\|y\| \leq$$

$$\leq [(1+\alpha)^{-1}KB^{1+\alpha}\|y\|^{\alpha} + CB]\|y\| \leq [(1+\alpha)^{-1}KB^{1+\alpha}r^{\alpha} + CB]\|y\| \leq q\|y\| ,$$

if $\|y\| \leq r$, where q is arbitrary with $BC < q < 1$ and $r^{\alpha} = (q-BC)(1+\alpha)/KB^{1+\alpha}$. Hence, by Lemma 3.3, Γ is a bounded directional contractor for P .

Theorem 3.3. Under the hypotheses of Lemma 3.4, P is a mapping onto Y

Proof. The proof follows immediately from Theorem 3.1 and Lemma 3.5.

Lemma 3.6. Let $P:X \to Y$ be a nonlinear operator differentiable in the Fréchet sense with Hölder continuous derivative $P'(x)$ satisfying (3.14). Let $\Gamma(x):Y \to X$ be a bounded linear operator and let C be a positive constant such that

$$\|P'(x)\Gamma(x)-I\| \leq C < 1 \quad \text{for all} \quad x \in X ,$$

where I is the identity mapping of Y. If, in addition,

$$\|\Gamma(x)\| \leq B \quad \text{for all} \quad x \in X ,$$

then Γ is a bounded directional contractor for P.

Proof. We have

$$\|P(x+\Gamma(x)y) - Px - y\| \leq \|P(x+\Gamma(x)y) - Px - P'(x)\Gamma(x)y\| +$$

$$+ \|P'(x)\Gamma(x)y - y\| \leq (1+\alpha)^{-1}K\|\Gamma(x)y\|^{1+\alpha} + C\|y\| \leq$$

$$\leq (1+\alpha)^{-1}KB^{1+\alpha}\|y\|^{1+\alpha} + C\|y\| \leq [(1+\alpha)^{-1}KB^{1+\alpha}\|y\|^{\alpha}+C]\|y\| \leq$$

$$\leq [(1+\alpha)^{-1}KB^{1+\alpha}r^{\alpha}+C]\|y\| \leq q\|y\| , \quad \text{if} \quad \|y\| \leq r ,$$

where q is arbitrary with $C < q < 1$ and $r^{\alpha} = (q-C)(1+\alpha)/KB^{1+\alpha}$. Hence, by Lemma 3.3, Γ is a bounded directional contractor for P.

Theorem 3.4. Under the hypotheses of Lemma 3.6, P is a mapping onto Y.

Proof. The proof follows immediately from Theorem 3.1 and Lemma 3.6.

4. A fixed point theorem

For operators $P = I - F$, where $Y = X$ and I is the identity mapping of X, it is convenient to have contractors of the form $I + \Gamma(x)$. Then the contractor inequality (2.1) becomes

$$(4.1) \qquad \|F(x+\epsilon(y+\Gamma(x)y)) - Fx - \epsilon\Gamma(x)y\| \leq q\epsilon\|y\|$$

for some $0 < \epsilon = \epsilon(x,y) \leq 1$, all $y \in X$ and $x \in D(F)$.

Thus, Theorem 3.1 yields immediately

Theorem 4.1. A closed nonlinear operator $F:D(F) \subset X \to X$ which has a bounded directional contractor, i.e., satisfying condition (4.1) and

$$\|\Gamma(x)\| \leq B \quad \text{for} \quad x \in D(F)$$

has a fixed point x^*, i.e., $x^* = Fx^*$. Moreover, $I-F$ is a mapping onto X.

This theorem also generalizes the well-known Banach fixed point theorem. In fact, if $F:X \to X$ is a contraction with Lipschitz constant $q < 1$, then $I+\Gamma(x)$ with $\Gamma(x) \equiv 0$ is obviously a bounded contractor (see Example 3.4, Chapter 1) and this notion is much stronger than a directional contractor. However, since the hypotheses of Theorem 4.1 are rather weak, we cannot prove the existence of the inverse mapping of $I-F$.

5. Evolution equations

We shall apply Theorem 4.1 to prove an existence theorem for nonlinear evolution equations.

Consider the initial value problem

$$(5.1) \qquad \frac{dx}{dt} = F(t,x), \quad 0 \leq t \leq T, \quad x(0) = \xi,$$

where $x = x(t)$ is a function defined on the real interval $[0,T]$ with values in the Banach space X, and $F:[0,T] \times X \to X$. Denote by X_T the space of all continuous functions $x = x(t)$ defined on $[0,T]$ with values in X and with the norm $\|x\|_C = \max[\|x(t)\| : 0 \leq t \leq T]$. Instead of (5.1) we consider the integral equation

$$(5.2) \qquad x(t) - \int_0^t F(s,x(s))ds = \xi$$

as an operator equation in X_T and we assume that the integral operator is closed in X_T.

For arbitrary fixed $x \in X$ and $t \in [0,T]$, let $\Gamma(t,x): X \to X$ be a bounded linear operator, strongly continuous with respect to (t,x) in the sense of the operator norm. Suppose that there exist positive numbers K and B such that the inequality

$$\max_{0 \le t \le T} \left\|F(t,x(t)+\rho\int_0^t \Gamma(s,x(s))y(s)ds) - F(t,x(t)) - \epsilon\Gamma(t,x(t))y(t)\right\| \le K\epsilon\|y\|_C$$

is satisfied for arbitrary continuous functions $x = x(t)$, $y = y(t) \in X_T$ and some $\epsilon = \epsilon(x,y) \le 1$, where

$$\|\Gamma(t,x)\| \le B \quad \text{for all} \quad x \in X \quad \text{and} \quad t \in [0,T] .$$

Then we say that $F(t,x)$ has a bounded directional contractor $\{I+\int_0^t \Gamma\}$ of integral type.

<u>Theorem 5.1</u>. Suppose that $F(t,x)$ has a bounded directional contractor satisfying (5.3) and T is such that $TK = q < 1$. Then for arbitrary $\xi \in X$ equation (5.2) has a continuous solution $x(t)$.

<u>Proof</u>. Following the method of proof of Theorem 3.1, we construct well-ordered sequences of numbers σ_α and elements x_α, $y_\alpha \in X_T$. Put $\sigma_0 = 0$, $x_0 = x_0(t) = \xi$ for $t \in [0,T]$ and $y_0 = y_0(t) =$ $= x_0(t) - \int_0^t F(s,x_0(s))ds - \xi$. Suppose that σ_γ, x_γ and y_γ have been constructed for all $\gamma < \alpha$, provided that, for arbitrary ordinal numbers $\gamma < \alpha$, inequality

$$(5.4_\gamma) \qquad \|y_\gamma\|_C \le e^{-(1-q)\sigma_\gamma}\|y_0\|_C ,$$

is satisfied, and for first kind ordinal numbers $\beta = \gamma+1 < \alpha$, the following inequalities are satisfied,

$$(5.5_{\gamma+1}) \qquad \|x_{\gamma+1}-x_\gamma\|_C \le (1+BT)e^{-(1-q)\sigma_\gamma}(\sigma_{\gamma+1}-\sigma_\gamma)\|y_0\|_C ,$$

$$(5.6_{\gamma+1}) \qquad \|y_{\gamma+1}-y_\gamma\|_C \le (1+q)e^{-(1-q)\sigma_\gamma}(\sigma_{\gamma+1}-\sigma_\gamma)\|y_0\|_C ,$$

$$(5.7_{\gamma+1}) \qquad 0 < \rho_\alpha = \sigma_{\alpha+1} - \sigma_\alpha < (1-q)^{-1}\ell n(1-q)(1-\bar{q})^{-1} ,$$

where $\bar{q}$ is arbitrary with $TK = q < \bar{q} < 1$; and for second kind (limit) numbers $\gamma < \alpha$ the following relations hold,

$$(5.8_\gamma) \qquad \sigma_\gamma = \lim_{\beta \nearrow \gamma} \sigma_\beta , \quad x_\gamma = \lim_{\beta \nearrow \gamma} x_\beta , \quad y_\gamma = \lim_{\beta \nearrow \gamma} y_\beta .$$

Then, in the same way as in the proof of Theorem 3.1, it follows from (5.5) - (5.8), Lemmas 3.1 and 3.2 that, for arbitrary $\gamma < \alpha$ and $\lambda < \alpha$, we have

$$(5.9) \qquad \|x_\gamma - x_\lambda\|_C \le (1-q)(1-\bar{q})^{-1}(1+BT)\int_{\sigma_\lambda}^{\sigma_\gamma} e^{-(1-q)\sigma} d\sigma \|y_0\|_C ,$$

$$(5.10) \qquad \|y_\gamma - y_\lambda\|_C \le (1-q^2)(1-\bar{q})^{-1}\int_{\sigma_\lambda}^{\sigma_\gamma} e^{-(1-q)\sigma} d\sigma \|y_0\|_C .$$

Suppose that $\beta = \alpha+1$ is a first kind ordinal number. Then we put

$$(5.11) \quad \sigma_{\alpha+1} = \sigma_\alpha + \rho_\alpha ; x_{\alpha+1}(t) = x_\alpha(t) - \rho_\alpha[y_\alpha(t) + \int_0^t \Gamma(s,x_\alpha(s))y_\alpha(s)ds] ,$$

$$(5.12) \qquad y_\alpha(t) = x_\alpha(t) - \int_0^t F(s,x_\alpha(s))ds - \xi, \ t \in [0,T] ,$$

where ρ_α is chosen so as to satisfy (5.7) and (5.3) with $\epsilon = \rho_\alpha$, $x = x_\alpha$ and $y = y_\alpha$, i.e.,

$$(5.13) \qquad \max_{0\le t\le T} \|F(t,x_\alpha(t) + \rho_\alpha \int_0^t \Gamma(s,x_\alpha(s))y_\alpha(s)ds) - F(t,x_\alpha(t)) - \rho_\alpha\Gamma(t,x_\alpha(t))y_\alpha(t)\| \le K\rho\|y_\alpha\|_C .$$

If $y_\alpha = 0$, then the proof of the theorem is completed. Otherwise, we have

$$y_{\alpha+1}(t) - y_\alpha(t) + \rho_\alpha y_\alpha(t) = = -\int_0^t [F(s,x_{\alpha+1}(s)) - F(s,x_\alpha(s)) + \rho_\alpha\Gamma(s,x_\alpha(s))y_\alpha(s)]ds .$$

Hence, it follows, by (5.11) and (5.12) with $-y_\alpha$ replacing y_α , that

(5.14) $$\|y_{\alpha+1}-y_\alpha+\rho_\alpha y_\alpha\|_C \leq q\rho_\alpha\|y_\alpha\|_C .$$

Hence,

$$\|y_{\alpha+1}\|_C \leq (1-(1-q)\rho_\alpha)\|y_\alpha\|_C < e^{-(1-q)\rho_\alpha}\|y_\alpha\|_C \leq e^{-(1-q)\sigma_{\alpha+1}}\|y_0\|_C ,$$

by virtue of (5.4_α) and (5.11). Thus, (5.4_α) implies $(5.4_{\alpha+1})$. Further, we obtain from (5.11) and (5.4_α)

$$\|x_{\alpha+1}-x_\alpha\|_C \leq (1+BT)\rho_\alpha\|y_\alpha\|_C \leq (1+BT)e^{-(1-q)\sigma_\alpha}(\sigma_{\alpha+1}-\sigma_\alpha)\|y_\alpha\|_C ,$$

that is, $(5.5_{\alpha+1})$ is satisfied. Now, we have

$$\|y_{\alpha+1}-y_\alpha\|_C \leq (1+q)\rho_\alpha\|y_\alpha\|_C \leq (1+q)e^{-(1-q)\sigma_\alpha}(\sigma_{\alpha+1}-\sigma_\alpha)\|y_0\|_C ,$$

by (5.14) and (5.4_α). Thus, conditions $(5.4_{\alpha+1})$, $(5.5_{\alpha+1})$ and $(5.6_{\alpha+1})$ are satisfied for $\sigma_{\alpha+1}$, $x_{\alpha+1}$ and $y_{\alpha+1}$.

Now suppose that α is an ordinal number of second kind and put $t_\alpha = \lim_{\gamma\nearrow\alpha} t_\gamma$. Let $\{\gamma_n\}$ be an increasing sequence convergent to α. It follows from (5.9) that $\|x_{\gamma_{n+p}}-x_{\gamma_n}\|_C \to 0$ as $n \to \infty$. Hence, the sequence $\{x_{\gamma_n}\}$ has a limit x_α and so does $\{x_\gamma\}$. It follows from (5.10) that the sequence $\{y_{\gamma_n}\}$ has a limit y_α and so does $\{y_\gamma\}$. Since the integral operator in (5.2) is closed in X_T, by assumption, we conclude that y_α satisfies (5.12). If $t_\alpha < +\infty$, then the limit passage in (5.4_{γ_n}) yields (5.4_α). The relations (5.8_α) are also satisfied by the definition of t_α, x_α and y_α and since we proved that (5.12) holds for y_α. The process will terminate if $\sigma_\alpha = +\infty$, where α is of second kind. In this case (5.4_α) yields $y_\alpha = 0$, i.e., x_α is a solution of (5.2) and the proof is completed.

Remark 5.1. It is not necessary that $F(t,x)$ be defined on the whole of X. It is sufficient to assume that $F(t,x)$ is defined for $x \in D$, where $D \subset X$ is a vector space. Then we assume, in addition, that $\Gamma(t,x)(X) \subset D$ for each $x \in D$ and $t \in [0,T]$.

Remark 5.2. Let us replace in Theorem 5.1 inequality (5.3) by the following one,

$$\|F(t,x(t)+\rho\int_0^t \Gamma(s,x(s))y(s)ds) - F(t,x(t)) - \rho\Gamma(t,x(t))y(t)\| \leq K\rho\|y(t)\| , \tag{5.15}$$

where $x = x(t)$, $y = y(t) \in X_T$ are arbitrary continuous functions and $0 < \rho = \rho(x,y) \leq 1$. Then one can prove the existence of a solution of the initial value problem (5.1) in the whole interval $[0,T]$.

In fact, following Bielecki [1], we define the norm $\|x\|_T$ of any function x in X_T to be

$$\|x\|_T = \sup_{0\leq t\leq T} e^{-Lt}\|x(t)\| ,$$

where L is a fixed positive constant greater than K . Then we replace the norm $\|\cdot\|_C$ by $\|\cdot\|_T$ everywhere in (5.4) - (5.7) and (5.9), (5.10). Further, instead of (5.13), we obtain

$$\|F(t,x_\alpha(t)+\rho_\alpha\int_0^t \Gamma(s,x_\alpha(s))y_\alpha(s)ds) - F(t,x_\alpha(t)) - \rho_\alpha\Gamma(t,x_\alpha(t))y_\alpha(t)\| \leq K\rho_\alpha\|y_\alpha(t)\| . \tag{5.16}$$

We have the relationship

$$y_{\alpha+1}(t) - y_\alpha(t) + \rho_\alpha y_\alpha(t) = -\int_0^t [F(s,x_{\alpha+1}(s)) - F(s,x_\alpha(s)) + \rho_\alpha\Gamma(s,x_\alpha(s))y_\alpha(s)]ds .$$

Hence, we obtain, by (5.16),

$$\|y_{\alpha+1}(t) - y_\alpha(t) + \rho_\alpha y_\alpha(t)\| =$$

$$= \|\int_0^t e^{Lt}e^{-L(t-s)}e^{-Ls}[F(s,x_{\alpha+1}(s)) - F(s,x_\alpha(s)) + \rho_\alpha\Gamma(s,x_\alpha(s))y_\alpha(s)]ds\| \leq$$

$$\leq e^{Lt}K\rho_\alpha \sup_{0\leq s\leq T} e^{-Ls}\|y_\alpha(s)\| \|\int_0^t e^{-L(t-s)}ds \leq e^{Lt}KL^{-1}\rho_\alpha\|y_\alpha\|_T .$$

Hence, it follows that

$$\|y_{\alpha+1}-y_\alpha+\rho_\alpha y_\alpha\|_T \leq q\rho_\alpha\|y_\alpha\|_T ,$$

where $q = KL^{-1} < 1$. Now, we can continue the same argument as in the proof of Theorem 5.1, where the norm $\|\cdot\|_C$ should be replaced by $\|\cdot\|_T$.

6. A local existence theorem

Using the directional contractor method we can prove a local existence theorem for solving nonlinear equations. Let X_0 be a vector space contained in the Banach space X . Put $S = S(x_0,r) = [x:\|x-x_0\| < r]$ for a given $x_0 \in X_0$ and $U = X_0 \cap \overline{S}$, where $\overline{S}$ is the closure of S . Let $P:U \to Y$ be a nonlinear operator closed on U . Suppose that P has a directional contractor $\Gamma(x):Y \to X$ for $x \in U_0 = X_0 \cap S$, i.e., there exists a positive constant $q < 1$ such that

$$\|P(x+\epsilon\Gamma(x)y) - Px - \epsilon y\| \leq q\epsilon\|y\| \tag{6.1}$$

for $x \in U_0$ and some $0 < \epsilon = \epsilon(x,y) \leq 1$, y being an arbitrary element of the Banach space Y .

Theorem 6.1. Suppose that the following hypotheses are satisfied,

1) P is closed on U

2) P has a bounded directional contractor $\Gamma:U_0 \to L(Y \to X)$ satisfying condition (6.1) and

3) $\|\Gamma(x)\| \leq B$ for $x \in U_0$

4) $r \geq B(1-\overline{q})^{-1}\|Px_0\|$, $q < \overline{q} < 1$.

Then equation $Px = 0$ has a solution in U .

Proof. In the same way as in the proof of Theorem 3.1, we construct the sequences of numbers $t_\alpha(t_0 = 0)$ and elements $x_\alpha \in U_0$ satisfying conditions (3.1_γ) for all ordinal numbers $\gamma < \alpha$ and conditions

$(3.2_{\gamma+1}) - (3.5_{\gamma+1})$ for first kind ordinal numbers $\beta < \alpha$.
Then using the same argument as in the proof of Theorem 3.1, we obtain for arbitrary $\gamma < \alpha$ and $\lambda < \alpha$

$$\|x_\gamma - x_\lambda\| \leq (1-q)(1-\bar{q})^{-1}B\|Px_0\|\int_{t_\lambda}^{t_\gamma} e^{-(1-q)t}dt .$$

Hence, we obtain

$$\|x_\gamma - x_0\| < (1-q)(1-\bar{q})^{-1}B\|Px_0\|\int_0^\infty e^{-(1-q)t}dt = B(1-\bar{q})^{-1}\|Px_0\| .$$

Hence, it follows from 3) that all elements x_γ belong to U_0 .

The further reasoning is exactly the same as in the proof of Theorem 3.1.

As a particular case of Theorem 6.1 we obtain

Theorem 6.2. If P is closed on U and has an inverse Gâteaux derivative $\Gamma(x)$, $x \in U_0$, satisfying conditions 3) and 4), then $Px = 0$ has a solution in U .

The proof follows from the fact that an inverse Gâteaux derivative is a directional contractor and there always exist $0 < q < 1/2$ satisfying condition (6.1). This theorem has been proved by Gavurin [1], where $\Gamma(x) = [P'(x)]^{-1}$, $x \in U_0$, $P'(x)$ being the linear Gâteaux derivative defined on X_0 and such that its inverse exists and is continuous on Y . In this case, $\Gamma(x)$ is an inverse Gâteaux derivative.

7. An implicit function theorem by means of directional contractors

Using Theorem 6.1 as a basis, an implicit function theorem can be proved. Let X, Z and Y be Banach spaces and put

$$S = [(x,z):\|x-x_0\| \leq r , \|z-z_0\| \leq \rho , x \in X , z \in Z]$$

for given $x_0 \in X$, $z_0 \in Z$, r and ρ .

Let $P:S \to Y$ be a continuous nonlinear operator and suppose that for every z such that $(x,z) \in S$, P has a directional contractor

$\Gamma(x,z):Y \to X$ which is z-uniform, i.e., there exists a positive $q < 1$ such that

(7.1) $$\|P(x+\epsilon\Gamma(x,z)y,z) - P(x,z) - \epsilon y\| \leq q\epsilon\|y\|$$

for some $0 < \epsilon \leq 1$, where y is an arbitrary element of Y. We assume, in addition, that the directional contractor $\Gamma(x,z)$ is bounded, i.e., there exists a constant B such that

(7.2) $$\|\Gamma(x,z)\| \leq B \quad \text{for} \quad (x,z) \in S .$$

Finally, we suppose that the directional contractor $\Gamma(x,z)$ is strongly continuous in (x,z), i.e., in the sense of the operator norm.

<u>Theorem 7.1</u>. Suppose that $P:S \to Y$ is a continuous operator satisfying the following conditions:

1) $P(x_0,z_0) = 0$.

2) P has a bounded directional z-uniform contractor satisfying conditions (7.1) and (7.2) and being strongly continuous with respect to (x,z). Then there exists a continuous function $g(z)$ defined in some neighborhood of z_0, with values in X, and such that $P(g(z),z) = 0$.

<u>Proof</u>. First of all we choose η and ρ_1 such that

(7.3) $$B(1-2q)^{-1}\eta \leq r \quad \text{and} \quad \|P(x_0,z)\| \leq \eta$$

for $\|z-z_0\| \leq \rho_1$. Now, in the same way as in the proof of Theorem 6.1, we construct sequences of numbers $t_\alpha (t_0 = 0)$ and continuous functions $x_\alpha(z)$ (replacing x_α), where $(x_\alpha,z) \in S_1$,

$$S_1 = [(x,z):\|x-x_0\| \leq r, \|z-z_0\| \leq \rho_1] \subset X \times Z .$$

The values of $x_\alpha(z)$ are in X and $x_0(z) \equiv x_0$. These sequences are to satisfy conditions (3.1_γ), $(3.2_{\gamma+1})$ - $(3.4_{\gamma+1})$ and $(6.2_{\gamma+1})$ provided $x_{\gamma+1}$, x_γ, $Px_{\gamma+1}$, Px_γ and Px_0 are replaced by $x_{\gamma+1}(z)$, $x_\gamma(z)$, $P(x_{\gamma+1}(z),z)$, $P(x_\gamma(z),z)$ and $P(x_0,z)$, respectively. Thus, we obtain in place of (3.5) and (3.6), by (7.3),

$$\|x_\gamma(z) - x_\lambda(z)\| \leq B\,\eta \int_{t_\lambda}^{t_\gamma} e^{-(1-q)t}\,dt \tag{7.4}$$

$$\|P(x_\gamma(z),z) - P(x_\lambda(z),z)\| \leq (1+q)\,\eta \int_{t_\lambda}^{t_\gamma} e^{-(1-q)t}\,dt\ . \tag{7.5}$$

Inequalities (7.4) and (7.5) show the z-uniform convergence in the relations which are analogous to (3.4_γ) yielding the continuity of the limit functions. An equivalent of (6.3) will be,

$$\|x_\gamma(z)-x_0\| \leq B(1-2q)^{-1}\eta \leq r\ ,$$

by virtue of (7.3). Hence, $(x_\gamma(z),z) \in S$. Thus, all constructed functions $x_\gamma(z)$ are continuous and well-defined. The further reasoning is exactly the same as in the proof of Theorem 6.1.

Since the assumption of Theorem 7.1 are rather weak, we cannot prove the uniqueness of the function $g(z)$.

CHAPTER 5

CONTRACTOR DIRECTIONS FOR SOLVING EQUATIONS

Introduction. Several authors have recently studied the solvability of nonlinear operator equations. Their main objective was to investigate a class of nonlinear operators having closed ranges. In Section 1, a generalization of some of the known results is presented and, in addition, new proofs are given.

In Section 2, a class of operators is discussed without assumption that the operators in question have closed ranges. In both sections, the technique of proofs is based on the concept of contractor directions. This concept is closely related to the notion of directional contractors and generalizes the concept of asymptotic directions introduced by Browder. A perturbation problem is discussed in Section 3. Finally, a generalization of contraction mappings, directional contractions, are discussed in Section 4 with application to evolution equations in Section 5.

1. Solvability for operators with closed range

Definition 1.1. Let X be an abstract set and $P:X \to Y$ a mapping of X into a (real or complex) Banach space Y, x a point in X. Then we define sets $\Gamma_x(P)$ of contractor directions for P at x, and $\Gamma_x(P)$ is such a set if there exists a positive constant $q = q(P) < 1$ with the following property: for each $y \in \Gamma_x(P)$, there exist a positive number $\epsilon = \epsilon(x,y) \leq 1$ and an element $\bar{x} \in X$ such that

$$\|P\bar{x}-Px-\epsilon y\| \leq q\epsilon\|y\| . \tag{1.1}$$

Thus, $\Gamma_x(P) = \Gamma_x(P,q)$ depends on $q = q(P)$, but q does not depend on $x \in X$ and obviously, $\Gamma_x(P,q) \subset \Gamma_x(P,\bar{q})$ if $0 < q < \bar{q} < 1$. The notation $\Gamma_x(P)$ will often replace $\Gamma_x(P,q)$.

Lemma 1.1. The closure of the set $\Gamma_x(P,q)$ is contained in $\Gamma_x(P,\overline{q})$ if $0 < q < \overline{q} < 1$. If $\Gamma_x(P)$ is dense in some ball $S(0,r) = [y:y \in Y, \|y\| \leq r]$, then $\Gamma_x(P) = Y$.

Proof. Let $0 < q < \overline{q} < 1$ and let $y \in \Gamma_x(P,q)$, $\overline{x} \in X$ satisfy (1.1). If $v \in X$ is such that $\|v-y\| \leq \eta$, then

$$\|P\overline{x}-Px-\epsilon v\| \leq \|P\overline{x}-Px-\epsilon y\| + \epsilon\|v-y\| \leq q\epsilon\|y\| + \epsilon\eta \leq q\epsilon(\|v\|+\eta) +$$

$$+ \epsilon\eta \leq \overline{q}\epsilon\|v\| \quad \text{if} \quad \eta \leq (\overline{q}-q)\|v\|(1+q)^{-1} ,$$

by virtue of (1.1).

Lemma 1.2. Let $y_0 \in Y$ be such that for all $x \in X$, $y_0 - Px$ is in the closure of $\Gamma_x(P,q)$. Let $\overline{q}$ be arbitrary with $0 < q < \overline{q} < 1$. If y_0 is not on the boundary of $P(X)$, then there exists a ball $S(y_0,r) \subset Y$ such that $y-Px \in \Gamma_x(P,\overline{q})$ for all $y \in S(y_0,r)$ and $x \in X$.

Proof. By Lemma 1.1, $y_0-Px \in \Gamma_x(P,\overline{\overline{q}})$ for all $x \in X$, where $\overline{\overline{q}}$ is arbitrary with $0 < q < \overline{\overline{q}} < \overline{q} < 1$. Thus, it follows that

$$\|P\overline{x}-Px-\epsilon(y_0-Px)\| \leq \overline{\overline{q}}\epsilon\|y_0-Px\|$$

is satisfied for all $x \in X$, where $\overline{x} \in X$ is properly chosen for $x \in X$. Hence, we obtain

$$\|P\overline{x}-Px-\epsilon(y-Px)\| \leq \|P\overline{x}-Px-\epsilon(y_0-Px)\| + \epsilon\|y_0-y\| \leq \overline{\overline{q}}\epsilon\|y_0-Px\| +$$

$$+ \epsilon r \leq \overline{\overline{q}}\epsilon(\|y-Px\|+r) + \epsilon r \leq \overline{q}\epsilon\|y-Px\|$$

$$\text{for all} \quad y \in S(y_0,r) \ , \quad x \in X ,$$

where $\overline{q}$ is arbitrary with $q < \overline{\overline{q}} < \overline{q} < 1$, and r is chosen so as to satisfy

$$(\overline{\overline{q}}+1)r \leq (\overline{q}-\overline{\overline{q}})d \leq (\overline{q}-\overline{\overline{q}})\|y-Px\| ,$$

where $2d$ is the distance from y_0 to the closure of $P(X)$.

Theorem 1.1. Let X be an abstract set, Y a real or complex Banach space, $P:X \to Y$ a mapping of X into Y such that the range $P(X)$ is closed in Y. Suppose that $y_0 \in Y$, which is not on the boundary of $P(X)$, is such that for each x in X, the element $y_0 - Px$ belongs to the closure of a set $\Gamma_x(P)$ of contractor directions for P at x. Then there exists a ball $S(y_0,r) \subset Y$ with center y_0 and radius r such that the equation

$$(1.2) \qquad Px = y, \quad y \in S(y_0,r), \quad x \in X$$

has a solution. Moreover, $y \in P(X)$ iff $y - Px \in \Gamma_x(P)$ for all $x \in X$.

Proof. One can assume that y_0 is not in $P(X)$. Then, by Lemma 1.2, there exists a sphere $S(y_0,r)$ such that $y - Px \in \Gamma_x(P)$ for all $y \in S(y_0,r)$, $x \in X$, i.e., there exist positive numbers $q < 1$, $\epsilon = \epsilon(x,y) \leq 1$ and elements $\bar{x} \in X$ such that

$$(1.3) \qquad \|P\bar{x} - Px - \epsilon(y-Px)\| \leq q\epsilon\|y-Px\|$$

for all $y \in S(y_0,r)$, $x \in X$. For the sake of simplicity, we replace equation (1.2) by $Px = 0$ in which the same symbol P stands for the new mapping with values $Px-y$, and fixed $y \in S(y_0,r)$. Then inequality (1.3) yields

$$(1.4) \qquad \|P\bar{x} - Px + \epsilon Px\| \leq q\epsilon\|Px\|.$$

Now, we construct well-ordered sequences of numbers t_α and elements $P(t_\alpha) \in P(X)$ as follows.

Put $t_0 = 0$ and let x_0 be an arbitrary element of X and put $P(t_0) = Px_0 \in P(X)$. Suppose that t_γ and $P(t_\gamma)$ have been constructed for all $\gamma < \alpha$, provided, for arbitrary number $\gamma < \alpha$ inequality (1.5_γ) is satisfied.

$$(1.5_\gamma) \qquad \|P(t_\gamma)\| \leq e^{-(1-q)t_\gamma}\|P(t_0)\|, \quad P(t_\gamma) \in P(X);$$

for first kind numbers $\beta = \gamma + 1 < \alpha$ the following inequalities are satisfied,

$$(1.6_{\gamma+1}) \qquad \|P(t_{\gamma+1}) - P(t_\gamma)\| \leq (1+q)\|P(t_0)\| e^{-(1-q)t_\gamma}(t_{\gamma+1} - t_\gamma) \ ;$$

$$(1.7_{\gamma+1}) \qquad 0 < t_{\gamma+1} - t_\gamma \leq 1$$

and for second kind numbers $\gamma < \alpha$ the following relations hold,

$$(1.8_\gamma) \qquad t_\gamma = \lim_{\beta \nearrow \gamma} t_\beta \ , \quad P(t_\gamma) = \lim_{\beta \nearrow \gamma} P(t_\beta) \ .$$

Then it follows from (1.6) - (1.8), Lemmas 3.1 and 3.2, Chapter 4, that for arbitrary $\lambda < \gamma < \alpha$ we have

$$\|P(t_\gamma) - P(t_\lambda)\| \leq \sum_{\lambda \leq \beta < \gamma} \|P(t_{\beta+1}) - P(t_\beta)\| \leq$$

$$\leq (1+q)\|P(t_0)\| \sum_{\lambda \leq \beta < \gamma} e^{-(1-q)t_\beta}(t_{\beta+1} - t_\beta) =$$

$$= (1+q)\|P(t_0)\| \sum_{\lambda \leq \beta < \gamma} e^{(1-q)(t_{\beta+1} - t_\beta)} e^{-(1-q)t_{\beta+1}}(t_{\beta+1} - t_\beta) \leq$$

$$\leq (1+q)e^{1-q} \sum_{\lambda \leq \beta < \gamma} e^{-(1-q)t_{\beta+1}}(t_{\beta+1} - t_\beta) <$$

$$< (1+q)e^{1-q} \sum_{\lambda \leq \beta < \gamma} \int_{t_\beta}^{t_{\beta+1}} e^{-(1-q)t} dt = (1+q)e^{1-q} \int_{t_\lambda}^{t_\gamma} e^{-(1-q)t} dt \ .$$

Hence,

$$(1.9) \qquad \|P(t_\gamma) - P(t_\lambda)\| \leq (1+q)e^{1-q} \int_{t_\lambda}^{t_\gamma} e^{-(1-q)t} dt \ .$$

Suppose that α is a first kind number. If $P(t_{\alpha-1}) = 0$, then the proof of the theorem is completed, since $P(t_{\alpha-1}) \in P(X)$. Suppose now that $P(t_{\alpha-1}) \neq 0$, and let $x \in X$ be such that $Px = P(t_{\alpha-1})$. Then there exist a positive $\epsilon \leq 1$ and an element $\overline{x} \in X$ satisfying (1.4). Put $\tau_\alpha = \epsilon \leq 1$,

$$(1.10) \qquad t_\alpha = t_{\alpha-1} + \tau_\alpha \ , \quad P(t_\alpha) = P\overline{x} \ .$$

Then we obtain by using (1.4) and the induction assumption $(1.5_{\alpha-1})$,

$$\|P(t_\alpha)\| \le [1-(1-q)\tau_\alpha]\|P(t_{\alpha-1})\| < e^{-(1-q)\tau_\alpha}\|P(t_{\alpha-1})\| \le$$

$$\le e^{-(1-q)\tau_\alpha} e^{-(1-q)t_{\alpha-1}}\|P(t_0)\| = e^{-(1-q)t_\alpha}\|P(t_0)\| ,$$

by (1.10). By virtue of (1.4), (1.10), we obtain

$$\|P(t_\alpha)-P(t_{\alpha-1})\| \le (1+q)\tau_\alpha\|P(t_{\alpha-1})\| \le (1+q)\|P(t_0)\|e^{-(1-q)t_{\alpha-1}}(t_\alpha-t_{\alpha-1}) .$$

Thus, conditions (1.5_α) - (1.7_α) are satisfied for t_α.

Now suppose that α is a number of second kind and put $t_\alpha = \lim_{\gamma\nearrow\alpha} t_\gamma$. Let $\{\gamma_n\}$ be an increasing sequence convergent to α. It follows from (1.9) that $\{P(t_{\gamma_n})\}$ is a Cauchy sequence and so is $\{P(t_\gamma)\}$. Denote by $P(t_\alpha)$ its limit. Since $P(t_\gamma)$ satisfy (1.5_γ), it follows that $P(t_\alpha)$ satisfies (1.5_α). The process will terminate if $t_\alpha = +\infty$, where α is of second kind. In this case (1.5_α) yields $P(t_\alpha) = 0$ and the proof is completed, since the necessity part is evident.

As a consequence of the above theorem, we obtain

<u>Theorem 1.2</u>. Let X be an abstract set, Y a real or complex Banach space, $P:X \to Y$ a mapping of X into Y such that the range $P(X)$ is closed in Y. Condition a) is necessary and sufficient for $P(X) = Y$.

a) Suppose that, for each x in X, a set $\Gamma_x(P)$ of contractor directions is dense in $S_r = S(0,r) \subset Y$.

<u>Proof</u>. By Lemma 1.1, the set $\Gamma_x(P)$ contains the ball S_r. Hence, it follows that $\Gamma_x(P) = Y$. Let $y_0 \in Y$ be not in $P(X)$. Since $y_0 - Px$ is in $\Gamma_x(P)$ for every $x \in X$, it follows from Theorem 1.1 that the equation $Px = y_0$ has a solution. This contradiction shows that $P(X) = Y$. The necessity of a) is evident.

<u>Remark 1.1</u>. Both theorems with some changes are proved in Section 6, for both real, vector space X and Banach space Y. However, the

proof given here is different and eliminates the use of a theorem by Zabreiko and Krasnosel'skii [1]. The above result presents a generalization of a theorem by Browder [1] and of a theorem by Zabreiko and Krasnosel'skii [1], Kirk and Caristi [1], and yields, in addition, new proofs.

2. Solvability for operators with arbitrary ranges

The restriction that the range of the operator in question is closed seems to play an essential role in the previous discussion. However, by introducing the concept of special contractor directions the mentioned restriction can be removed.

Denote by $\mathbb{B}$ (see Chapter 3) the class of increasing continuous functions $B(s)$ such that $B(s) > 0$ for $s > 0$ and

$$\int_0^a s^{-1}B(s)\,ds < \infty \quad \text{for some} \quad a > 0 \,. \tag{2.1}$$

Definition 2.1. Let X be a complete metric space and let Y be a real or complex Banach space, $P:D(P) \subset X \to Y$, and $x \in X$. Then $\Gamma_x(P) = \Gamma_x(P,q)$ is a set of special contractor directions for P at $x \in D(P)$ if there exist a positive constant $q < 1$ and a function $B \in \mathbb{B}$ which have the following property. For each $y \in \Gamma_x(P)$, there exist a positive number $\epsilon = \epsilon(x,y) \leq 1$ and an element $\bar{x} \in D(P)$ such that

$$\|P\bar{x}-Px-\epsilon y\| \leq q\,\epsilon\,\|y\| \tag{2.2}$$

and

$$d(\bar{x},x) \leq \epsilon B(\|y\|) \,, \tag{2.3}$$

where $d(\cdot,\cdot)$ is the distance in X.

Thus, $\Gamma_x(P) = \Gamma_x(P,q)$ depends on $q = q(P)$, but q does not depend on $x \in D(P)$ and obviously, $\Gamma_x(P,q) \subset \Gamma_x(P,\bar{q})$ if $0 < q < \bar{q} < 1$. The notation $\Gamma_x(P)$ will often replace $\Gamma_x(P,q)$.

Lemma 2.1. The closure of the set $\Gamma_x(P,q)$ is contained in $\Gamma_x(P,\bar{q})$ if $0 < q < \bar{q} < 1$. If $\Gamma_x(P)$ is dense in some ball $S(0,r) =$

$= [y : y \in Y, \|y\| \le r]$, then $\Gamma_x(P) = Y$.

Proof. Let $0 < q < \overline{q} < 1$ and let $y \in \Gamma_x(P,q)$, $\overline{x} \in D(P)$ satisfy (1.2) and (1.3). If $v \in S(0,r)$ is such that $\|v-y\| \le \eta$, then

$$\|P\overline{x}-Px-\epsilon v\| \le \|P\overline{x}-Px-\epsilon y\| + \epsilon\|v-y\| \le q\epsilon\|y\| + \epsilon\eta \le q\epsilon(\|v\|+\eta) +$$

$$+ \epsilon\eta \le \overline{q}\epsilon\|v\| \quad \text{if} \quad \eta \le (\overline{q}-q)\|v\|(1+q)^{-1} ,$$

by virtue of (2.2).

Lemma 2.2. Let $y_0 \in Y$ be such that for all $x \in D(P)$, y_0-Px is in the closure of $\Gamma_x(P,q)$. Let $\overline{q}$ be arbitrary with $0 < q < \overline{q} < 1$. If y_0 is not on the boundary of $P(D(P))$, then there exists a sphere $S(y_0,r) \subset Y$ such that $y-Px \in \Gamma_x(P,\overline{q})$ for all $y \in S(y_0,r)$ and $x \in D(P)$.

Proof. By Lemma 1.1, $y_0-Px \in \Gamma_x(P,\overline{\overline{q}})$ for all $x \in D(P)$, where $\overline{\overline{q}}$ is arbitrary with $0 < q < \overline{\overline{q}} < \overline{q} < 1$. Thus, it follows that

$$\|P\overline{x}-Px-\epsilon(y_0-Px)\| \le \overline{\overline{q}}\epsilon\|y_0-Px\|$$

is satisfied for all $x \in D(P)$, where $\overline{x} \in D(P)$ is properly chosen for $x \in D(P)$ so as to satisfy (2.2). Hence, we obtain

$$\|P\overline{x}-Px-\epsilon(y-Px)\| \le \|P\overline{x}-Px-\epsilon(y_0-Px)\| + \epsilon\|y_0-y\| \le \overline{\overline{q}}\epsilon\|y_0-Px\| +$$

$$+ \epsilon r \le \overline{\overline{q}}\epsilon(\|y-Px\|+r) + \epsilon r \le \overline{q}\epsilon\|y-Px\|$$

$$\text{for all } y \in S(y_0,r) , \quad x \in D(P) ,$$

where $\overline{q}$ is arbitrary with $q < \overline{\overline{q}} < \overline{q} < 1$, and r is chosen so as to satisfy

$$(\overline{\overline{q}}+1)r \le (\overline{q}-\overline{\overline{q}})d \le (\overline{q}-\overline{\overline{q}})\|y-Px\| ,$$

where $2d$ is the distance from y_0 to the closure of $P(D(P))$.

Theorem 2.1. Let $P : D(P) \subset X \to Y$ be a closed operator, Y being a real or complex Banach space. Suppose that $y_0 \in Y$ is such that, for each $x \in D(P)$, the element $y_0 - Px$ belongs to the closure of a set $\Gamma_x(P)$ of contractor directions defined by means of (2.2) and (2.3). Then the equation

$$Px - y_0 = 0 \ , \quad x \in D(P)$$

has a solution. If, in addition, y_0 is not a limit point of $P(D(P))$, then there exists a ball $S(y_0, r)$ such that the equation $Px - y = 0$ has a solution for each $y \in S(y_0, r)$.

Proof. Although the proof is similar to that of Theorem 1.1, it will be useful to supply here a complete proof, because of a new sequence involved in the induction procedure. Without loss of generality, assume that $y_0 = 0$. We construct well-ordered sequences of positive numbers t_α and elements $x_\alpha \in D(P)$ as follows. Put $t_0 = 0$, and let x_0 be an arbitrary element of $D(P)$. Suppose that t_γ and x_γ have been constructed for all $\gamma < \alpha$, provided, for arbitrary ordinal numbers $\gamma < \alpha$, inequality

$$(2.4_\gamma) \qquad \|Px_\gamma\| \le e^{-(1-q)t_\gamma} \|Px_0\|$$

is satisfied and for first kind ordinal numbers, $\beta = \gamma + 1 < \alpha$, the following inequalities are satisfied,

$$(2.5_{\gamma+1}) \qquad 0 < t_{\gamma+1} - t_\gamma \le 1 \ ,$$

$$(2.6_{\gamma+1}) \qquad d(x_{\gamma+1}, x_\gamma) \le B(\|Px_0\| e^{-(1-q)t_\gamma})(t_{\gamma+1} - t_\gamma) \ ,$$

$$(2.7_{\gamma+1}) \qquad \|Px_{\gamma+1} - Px_\gamma\| \le (1+q)\|Px_0\| e^{-(1-q)t_\gamma}(t_{\gamma+1} - t_\gamma) \ ;$$

and, for second kind (limit) ordinal numbers, $\gamma < \alpha$, the following relations hold,

$$(2.8_\gamma) \qquad t_\gamma = \lim_{\beta \nearrow \gamma} t_\beta \ , \quad x_\gamma = \lim_{\beta \nearrow \gamma} x_\beta \ , \quad Px_\gamma = \lim_{\beta \nearrow \gamma} Px_\beta \ .$$

Then it follows from (2.6), (2.8), Lemmas 3.1 and 3.2, Chapter 4, that, for arbitrary $\lambda < \gamma < \alpha$, we have

$$d(x_\gamma,x_\lambda) \le \sum_{\lambda\le\beta<\gamma} d(x_{\beta+1},x_\beta) \le \sum_{\lambda\le\beta<\gamma} B(\|Px_0\|e^{-(1-q)t_\beta})(t_{\beta+1}-t_\beta) =$$

$$= \sum_{\lambda\le\beta<\gamma} B(\|Px_0\|e^{(1-q)(t_{\beta+1}-t_\beta)}e^{-(1-q)t_{\beta+1}})(t_{\beta+1}-t_\beta) <$$

$$< \sum_{\lambda\le\beta<\gamma} B(e^{1-q}\|Px_0\|e^{-(1-q)t_{\beta+1}})(t_{\beta+1}-t_\beta) <$$

$$< \sum_{\lambda\le\beta<\gamma} \int_{t_\beta}^{t_{\beta+1}} B(e^{1-q}\|Px_0\|e^{-(1-q)t}\,dt = \int_{t_\lambda}^{t_\gamma} B(e^{1-q}\|Px_0\|e^{-(1-q)t})\,dt \ .$$

Hence, we obtain the following estimate,

$$d(x_\gamma,x_\lambda) \le \int_{t_\lambda}^{t_\gamma} B(e^{1-q}\|Px_0\|e^{-(1-q)t})\,dt \ . \tag{2.9}$$

In the same way we obtain from (2.5), (2.7), Lemmas 3.1 and 3.2, Chapter 4, that

$$\|Px_\gamma-Px_\lambda\| \le (1+q)e^{1-q}\|Px_0\| \int_{t_\lambda}^{t_\gamma} e^{-(1-q)t}dt \ . \tag{2.10}$$

Suppose that α is a first kind ordinal number. If $Px_{\alpha-1} = 0$, then the proof of the first statement of the theorem is completed.

If $Px_{\alpha-1} \ne 0$, then we put

$$t_\alpha = t_{\alpha-1} + \tau_\alpha \ , \quad x_\alpha = \overline{x} \ , \tag{2.11}$$

where $\overline{x} \in D(P)$ and $\tau_\alpha = \epsilon \le 1$ are chosen so as to satisfy (2.2) and (2.3) with $x = x_{\alpha-1}$, $y = y_0-Px_{\alpha-1} = -Px_{\alpha-1}$, since $y_0-Px_{\alpha-1} \in \Gamma_x(P,\overline{q})$ is a contractor direction for P at $x = x_{\alpha-1}$, by virtue of Lemma 2.1, $\overline{q}$ being arbitrary with $q < \overline{q} < 1$. Thus, we have (writing q instead of $\overline{q}$),

$$\|Px_\alpha-Px_{\alpha-1}+\tau_\alpha Px_{\alpha-1}\| \le q\tau_\alpha\|Px_{\alpha-1}\| \ . \tag{2.12}$$

Hence, we obtain, by (2.12) and $(2.4_{\alpha-1})$,

$$\|Px_\alpha\| \leq (1-\tau_\alpha)\|Px_{\alpha-1}\| + q\tau_\alpha\|Px_{\alpha-1}\| = (1-(1-q)\tau_\alpha)\|Px_{\alpha-1}\| <$$

$$< e^{-(1-q)\tau_\alpha}\|Px_{\alpha-1}\| \leq e^{-(1-q)t_\alpha}\|Px_0\| ,$$

i.e.,

$$(2.13_\alpha) \qquad \|Px_\alpha\| \leq e^{-(1-q)t_\alpha}\|Px_0\| .$$

It follows from (2.11), (2.3) and $(2.4_{\alpha-1})$ that

$$(2.14_\alpha) \quad d(x_\alpha, x_{\alpha-1}) \leq \tau_\alpha B(\|Px_{\alpha-1}\|) \leq (t_\alpha - t_{\alpha-1})B(\|Px_0\|e^{-(1-q)t_{\alpha-1}}) .$$

It follows from (2.11), (2.12) and $(2.4_{\alpha-1})$ that

$$(2.15_\alpha) \quad \|Px_\alpha - Px_{\alpha-1}\| \leq (1+q)\tau_\alpha\|Px_{\alpha-1}\| \leq (1+q)\|Px_0\|e^{-(1-q)t_{\alpha-1}}(t_\alpha - t_{\alpha-1}) .$$

Thus, relations (2.4_α) - (2.7_α) are also satisfied for t_α and x_α. Now suppose that α is an ordinal number of second kind and put $t_\alpha = \lim_{\gamma \nearrow \alpha} t_\gamma$. Let $\{\gamma_n\}$ be an increasing sequence convergent to α. It follows from (2.9) and (2.10) that $\{x_{\gamma_n}\}$ and $\{Px_{\gamma_n}\}$ are Cauchy sequences and so are $\{x_\gamma\}$ and $\{Px_\gamma\}$. Denote by x_α and y_α their limits, respectively. Since P is closed, we infer that $x_\alpha \in D(P)$ and $y_\alpha = Px_\alpha$. If $t_\alpha < \infty$, then the limit passage in (2.4_{γ_n}) yields (2.4_α). The relationships (2.8_α) are satisfied by the definition of t_α and x_α, since $y_\alpha = Px_\alpha$. This process will terminate if $t_\alpha = \infty$, where α is of second kind. In this case, $Px_\alpha = 0$, by virtue of (2.4_α). The limit x_α exists, by (2.9), since

$$(2.16) \qquad \int_0^\infty B(e^{1-q}\|Px_0\|e^{-(1-q)t})dt = (1-q)^{-1}\int_0^a s^{-1}B(s)ds < \infty ,$$

where $a = e^{1-q}\|Px_0\|$. Thus the proof of the first assertion of the theorem is completed. To prove the remaining assertion, we apply Lemma 1.2 and the proof then follows from the first assertion.

As a consequence of the above theorem, we obtain:

Theorem 2.2. Let $P:D(P) \subset X \to Y$ be a closed operator, where X is a complete metric space and Y is a real or complex Banach space. Suppose that for each $x \in D(P)$ a set $\Gamma_x(P)$ of special contractor directions for P is dense in some sphere with center 0 in Y. Then $P(D(P)) = Y$.

Proof. By Lemma 2.1, we have that $\Gamma_x(P) = Y$ for all $x \in D(P)$. The proof follows from the first assertion of Theorem 2.1, since $y_0 - Px \in \Gamma_x(P)$ for all $y_0 \in Y$ and $x \in D(P)$.

A local existence theorem

By using the method of contractor directions, a local existence theorem for nonlinear operator equations can also be proved. A similar local existence theorem is given above under more restrictive conditions for the directional contractor method. It turns out that the applicability of this theorem depends essentially on the restrictions imposed on the sets of contractor directions.

Let X_0 be a subset of the complete metric space X. Put $S = S(x_0, r) = [x : d(x, x_0) < r,\ x \in X]$ for a given $x_0 \in X_0$, and $U = X_0 \cap \bar{S}$, where $\bar{S}$ is the closure of S in X. Let $P:U \to Y$ be a nonlinear operator closed on U, i.e., $x_n \in U$, $x_n \to x$ and $Px_n \to y$ imply that $x \in U$ and $y = Px$, where y is an element of the Banach space Y.

Theorem 2.3. Suppose that the following hypotheses are satisfied:

1) $P:U \to Y$ is closed on U;
2) for each $x \in U_0 = X_0 \cap S$, a set $\Gamma_x(P)$ of special contractor directions is dense in some sphere with center in Y.
3) $r \geq (1-q)^{-1} \int_0^a s^{-1} B(s)\,ds$, where $a = (1-q)(1-\bar{q})^{-1} \|Px_0\|$, $\bar{q}$ is arbitrary with $q < \bar{q} < 1$, and q, $B(s)$ are defined by (2.2), (2.3) and (2.1), respectively.

Then equation $Px = 0$ has a solution x in U .

Proof. By Lemma 2.1, $\Gamma_x(P) = Y$ for all $x \in U_0$. Now as in the proof of Theorem 2.1, we construct the sequences $\{t_\alpha\}$ and $\{x_\alpha\}$ which satisfy the required induction assumptions and, in addition, instead of $(2.5_{\gamma+1})$,

$$0 < \tau_{\gamma+1} = t_{\gamma+1}-t_\gamma < (1-q)^{-1}\ell n(1-q)(1-\bar{q})^{-1} ,$$

for first kind ordinal numbers $\beta = \gamma+1 < \alpha$. Hence, we obtain the following estimate instead of (2.9),

$$d(x_\gamma,x_\lambda) \le \sum_{\lambda\le\beta<\gamma} B(\|Px_0\|e^{-(1-q)t_\beta})(t_{\beta+1}-t_\beta) =$$

$$= \sum_{\lambda\le\beta<\gamma} B(\|Px_0\|e^{(1-q)(t_{\beta+1}-t_\beta)}e^{-(1-q)t_{\beta+1}})(t_{\beta+1}-t_\beta) <$$

$$< \sum_{\lambda\le\beta<\gamma} B((1-q)(1-\bar{q})^{-1}\|Px_0\|e^{-(1-q)t_{\beta+1}})(t_{\beta+1}-t_\beta) <$$

$$< \sum_{\lambda\le\beta<\gamma} \int_{t_\beta}^{t_{\beta+1}} B((1-q)(1-\bar{q})^{-1}\|Px_0\|e^{-(1-q)t})dt =$$

$$= \int_{t_\lambda}^{t_\gamma} B((1-q)(1-\bar{q})^{-1}\|Px_0\|e^{-(1-q)t})dt .$$

Hence, we obtain

$$d(x_\gamma,x_\lambda) \le \int_{t_\lambda}^{t_\gamma} B((1-q)(1-\bar{q})^{-1}\|Px_0\|e^{-(1-q)t})dt , \tag{2.17}$$

and

$$\int_0^\infty B((1-q)(1-\bar{q})^{-1}\|Px_0\|e^{-(1-q)t})dt = (1-q)^{-1}\int_0^a s^{-1}B(s)ds , \tag{2.18}$$

where $a = (1-q)(1-\bar{q})^{-1}\|Px_0\|$.

In the same way we obtain instead of (2.10),

$$\|Px_\gamma-Px_\lambda\| \le (1-q^2)(1-\bar{q})^{-1}\|Px_0\|\int_{t_\lambda}^{t_\gamma} e^{-(1-q)t}dt .$$

It follows in particular from (2.17) and (2.18) that

$$d(x_\gamma, x_0) \leq (1-q)^{-1} \int_0^a s^{-1} B(s)\,ds \ .$$

The fact that the indicated replacement of $(2.5_{\gamma+1})$ is feasible, is based on the following remark.

Remark 2.1. If a set $\Gamma_x(P)$, $x \in D(P)$, of special contractor directions is dense in some ball with center 0 in Y , then, for arbitrary $\tau > 0$, $y \in Y$, there is a positive $\epsilon \leq \tau$ satisfying conditions (2.2) and (2.3). In fact, since $\overline{\Gamma_x(P)} = Y$, by Lemma 2.1, for τy , there is an $\epsilon_0 = \epsilon_0(x, \tau y) \leq 1$ which satisfies (2.2) and (2.3) with $\epsilon = \epsilon_0$ and y replaced by τy . Hence, (2.2) and (2.3) hold for $\epsilon = \epsilon_0 \tau \leq \tau$.

Remark 2.2. Condition 3) of Theorem 2.3 can be replaced by

$$\text{3a)} \qquad r \geq (1-q)^{-1} \int_0^a s^{-1} B(s)\,ds \ ,$$

where $a = e^{1-q}\|Px_0\|$. In this case, we use the estimates (2.9), (2.10), (2.16) and the induction assumption $(2.5_{\gamma+1})$.

A theorem on closed linear transformations

Let $A{:}D(P) \subset X \to Y$ be a closed linear operator, X and Y being Banach spaces. Suppose that there exist a positive constant $q < 1$ and a function $B \in \mathbb{B}$ and a dense set $V \subset Y$ which have the following property. For each $y \in V$, there exists an element h in the vector space $D(P)$ such that

$$\text{(2.19)} \qquad \|Ah-y\| \leq q\|y\| \quad \text{and} \quad \|h\| \leq B(\|y\|) \ .$$

Then A maps X onto Y .

Proof. We may assume that $V = Y$, since then (2.19) will be satisfied for Y and arbitrary $q < \bar{q} < 1$. Let $y_0 \in Y$ be arbitrary and put

$$y_{n+1} = Ah_n - y_n \quad \text{for} \quad n = 0,1,\cdots \ ,$$

where h_n is chosen so as to satisfy (2.19) with $h = h_n$ and $y = y_n$.

Then we have

$$\|y_n\| \leq q\|y_{n-1}\| \leq q^n\|y_0\| \quad \text{and} \quad y_{n+1} = A(h_n - h_{n-1} + \ldots + (-1)^n h_0) - (-1)^n y_0 .$$

But

$$\sum_{n=0}^{\infty} \|h_n\| \leq \sum_{n=0}^{\infty} B(\|y_n\|) \leq \sum_{n=0}^{\infty} B(q^n\|y_0\|) < \infty ,$$

by virtue of (2.19), (see Chapter 3). Hence, it follows that equation $Ah = y_0$ has a solution.

Remark 2.2. The necessity of condition (2.19) follows from a well-known theorem (see Theorem 2.12.1, Hille and Phillips [1]).

Lemma 2.1. Let $A:X \to Y$ be a bounded linear operator which maps the Banach space X onto the Banach space Y, and let $B > \|(A^*)^{-1}\|$, where A^* is the adjoint operator. Then, for each $y \in Y$, there exists $h \in X$ such that $Ah = y$ and $\|h\| \leq B\|y\|$.

Proof. Denote by θ the set of all solutions of the equation $Ax = 0$. The transformation A defines on the quotient space X/θ a bounded linear operator $\mathcal{A}:X/\theta \to Y$ by the formula $y = \mathcal{A}\xi$, where $x \in \xi \in X/\theta$ and $y = Ax$. By the lemma of Altman [17], $\|\mathcal{A}^{-1}\| = \|(A^*)^{-1}\|$ (the domain of $(A^*)^{-1}$ is not, in general, the whole space). Since $\|\xi\| \leq \|\mathcal{A}^{-1}\|\|y\|$, where $\|\xi\| = \inf[\|x\|:x \in \xi]$, there exists an element $x \in \xi$ such that $\|x\| \leq \|\xi\|(1+\eta)$, where η is determined by the relationship $\|\mathcal{A}^{-1}\|(1+\eta) = B$. Then we obtain

$$\|x\| \leq \|\xi\|(1+\eta) \leq \|\mathcal{A}^{-1}\|(1+\eta)\|y\| = B\|y\| ,$$

and $Ax = y$.

As an application of Theorem 2.2, we obtain

Theorem 2.4. Let $P:X \to Y$ be a closed mapping of the Banach space X into the Banach space Y. Suppose that P is differentiable in the Gâteaux sense and its derivative $P'(x)$ maps X onto Y for each $x \in X$. If there exists a constant B such that $\|(P'(x)^*)^{-1}\| \leq B$

for all $x \in X$, then $P(X) = Y$.

<u>Proof</u>. Let $y \in Y$ and q be arbitrary with $0 < q < 1$. By Lemma 2.1, there exists an element $h \in X$ such that

$$P'(x)h - y = 0 \quad \text{and} \quad \|h\| \leq \overline{B}\|y\| ,$$

where $\overline{B} > B$ is arbitrary. Choose a positive $\epsilon \leq 1$ such that

$$\|P(x+\epsilon h) - Px - \epsilon P'(x)h\| \leq q\epsilon\|y\| .$$

Then we obtain

$$\|P(x+\epsilon h) - Px - \epsilon y\| \leq \|P(x+\epsilon h) - Px - \epsilon P'(x)h\| + \epsilon\|P'(x)h - y\| \leq \epsilon q\|y\|$$

and $\|\overline{x}-x\| \leq \overline{B}\epsilon\|y\|$ is true for $\overline{x} = x+\epsilon h$, that is, condition (2.3) is satisfied. Hence, it follows that the set of special contractor directions $\Gamma_x(P)$ is the whole of Y for all $x \in X$ and Theorem 2.2 is applicable.

<u>A local existence theorem</u>

A local existence theorem can be obtained on the basis of Theorem 2.3.

Let X_0 be a vector space contained in the Banach space X , $S = S(x_0,r)$ for a given $x_0 \in X_0$ and $U = X_0 \cap \overline{S}$.

<u>Theorem 2.5</u>. Let $P:U \to Y$ be closed on U . For each $x \in U_0 = X_0 \cap S$, the Gâteaux derivative $P'(x)$ exists and maps X onto the Banach space Y . There exist positive constants $\overline{B}$, $\overline{q} < 1$ such that $r > \overline{B}(1-\overline{q})^{-1}\|Px_0\|$ and $\|(P'(x)^*)^{-1}\| \leq \overline{B}$ for all $x \in U_0$. Then equation $Px = 0$ has a solution in U .

<u>Proof</u>. There exists $B > \overline{B}$ such that $r \geq B(1-\overline{q})^{-1}\|Px_0\|$. By using the same argument as in the proof of Theorem 2.3, one can see that condition (2.3) is satisfied for arbitrary positive $q < \overline{q}$. Hence, it follows that the set $\Gamma_x(P)$ of special contractor directions is the whole of Y for all $x \in U_0$. Thus, the hypotheses of Theorem 2.3 are

satisfied.

3. Lipschitz approximation and a perturbation problem

Let X , Y be Banach spaces. A mapping $T:X \to Y$ is called a Lipschitz approximation to the mapping $P:X \to Y$ if $P-T$ is Lipschitz continuous, that is, there exists a constant C such that

$$\|(Px-Tx) - (P\bar{x}-T\bar{x})\| \leq C\|x-\bar{x}\| \quad \text{for all } x,\bar{x} \in X . \tag{3.1}$$

Lemma 3.1. Let T be a Lipschitz approximation to P . Suppose that there exist positive constants $q < 1$, B , and a ball S_r in Y and a set S which is dense in S_r such that, for every $y \in S$, there exists a positive number $\epsilon = \epsilon(x,y) \leq 1$ and an element $h = h(x,y) \in X$ satisfying the following inequality

$$\|T(x+h) - Tx - \epsilon y\| \leq q\epsilon\|y\| \quad \text{and} \quad \|h\| \leq B\|\epsilon y\| . \tag{3.2}$$

If $q + BC < 1$, then $\Gamma_x(P) = Y$.

Proof. Let $v \in S_r$ be arbitrary but fixed and let η , $\bar{q}$ be positive constants with $q + BC < \bar{q} < 1$. Choose $y \in S$ such that $\|y-v\| \leq \eta\|v\|$ and $h = h(x,y)$ satisfying (3.2).

Then we obtain

$$\|P(x+h) - Px - \epsilon v\| \leq \|T(x+h) - Tx - \epsilon v\| +$$

$$+ \|[P(x+h) - T(x+h)] - [Px-Tx]\| \leq q\epsilon\|y\| + \epsilon\|y-v\| + C\|h\| \leq$$

$$\leq q\epsilon\|y\| + \epsilon\eta\|v\| + BC\epsilon\|y\| \leq q\epsilon(1+\eta)\|v\| + \epsilon\eta\|v\| + BC\epsilon(1+\eta)\|v\| =$$

$$= \epsilon[q(1+\eta) + \eta + BC(1+\eta)]\|v\| \leq \bar{q}\epsilon\|v\| ,$$

if η is chosen so as to satisfy $\eta(q+BC+1) \leq \bar{q} - (q+BC)$. Hence, it follows that the set $\Gamma_x(P)$ of contractor directions for P at $x \in X$ contains the sphere S_r . Thus, $\Gamma_x(P) = Y$.

Theorem 3.1. Let $P:X \to Y$ be a closed mapping satisfying the hypotheses of Lemma 3.1. Then $P(X) = Y$.

Proof. The proof follows immediately from Theorem 2.2 and Lemma 3.1.

Theorem 3.2. Let $P:D(P) \subset X \to Y$ be a closed nonlinear operator, where X and Y are real or complex Banach spaces and $D(P)$ is a vector space. Suppose that, for each $x \in D(P)$, the set $\Gamma_x(P)$ of special contractor directions for P is dense in some ball with center 0 in Y. If $L:X \to Y$ is a Lipschitz continuous operator and $|\delta|$ is sufficiently small, then $P + \delta L$ is a mapping onto Y.

Proof. By assumption, there exist a ball S_r with center 0 in Y, a set S which is dense in S_r, and positive constants $q < 1$, B such that for every $y \in S$, there exist a positive number $\epsilon = \epsilon(x,y) \leq 1$ and an element $h = h(x,y) \in X$ satisfying the following inequalities,

$$\|P(x+h) - Px - \epsilon y\| \leq q\epsilon\|y\| \quad \text{and} \quad \|h\| \leq B\epsilon\|y\| .$$

But P is a Lipschitz approximation to $P + \delta L$ with Lipschitz constant $C = \delta C_L$, where C_L is the Lipschitz constant for the operator L. If δ is such that $q + |\delta| BC_L < 1$, then the hypotheses of Lemma 3.1 are satisfied with P and $P + \delta L$ replacing T and P, respectively. Thus, the proof follows from Theorem 3.1.

4. Directional contractions

A generalization of the notion of the contraction mapping in a Banach space can be given which is based on the concept of contractor direction.

Definition 4.1. Let X be a Banach space. A mapping $F:X \to X$ is called a directional contraction if there exists a positive number $q < 1$ which has the following property.

For arbitrary $x,y \in X$, there exists a positive number $\epsilon = \epsilon(x,y) \leq 1$ such that

$$\|F(x+\epsilon y) - Fx\| \leq q\epsilon\|y\| . \tag{4.1}$$

<u>Theorem 4.1</u>. Suppose that $F:X \to X$ is a directional contraction and P is a closed operator, where $Px = x - Fx$. Then P maps the Banach space X onto itself.

<u>Proof</u>. The proof follows from Theorem 2.2. In fact, if $y \in Y = X$ is arbitrary and ϵ is such that it satisfies condition (4.1), then (2.2) is satisfied with $\bar{x} = x + \epsilon y$. Hence, it follows that $\Gamma_x(P) = X$. The additional requirement that $\|\bar{x}-x\| \leq B\epsilon\|y\|$ is also satisfied with $B = 1$.

<u>Remark 4.1</u>. It is easy to show that Theorem 4.1 follows immediately from the basic directional contractor theorem (Theorem 3.1, Chapter 4). In fact, condition (4.1) coincides with the contractor inequality for $Px = x - Fx$ if $\Gamma(x) = I$ for all $x \in X$. This implies that the identity operator I is a bounded contractor for $P = I - F$.

The following fixed point theorem can be proved as a particular case of Theorem 2.3.

<u>Theorem 4.2</u>. Let S be an open ball with radius r and center θ (zero) in Banach space X and let $F:\bar{S} \to X$ be a directional contraction on S, i.e., for each $x \in S$ and $y \in Y$ with $\|y\| \leq r$, there exists a positive $\epsilon = \epsilon(x,y) \leq 1$ satisfying (4.1) and $x + \epsilon y \in S$. If P is closed, where $Px = x - Fx$ and $r > \|F\theta\|/(1-q)$, then there exists an element $x^* \in \bar{S}$ such that $x^* = Fx^*$.

<u>Proof</u>. It is easy to see that the hypotheses of Theorem 2.3 are satisfied with $x_0 = \theta$. Obviously, $\Gamma_x(P) = X$ and condition 3) of Theorem 2.3 that $r \geq B(1-\bar{q})^{-1}\|Px_0\|$ is satisfied with $B = 1$ and $\bar{q}$ being chosen so as to satisfy

$$q < \bar{q} < 1 \quad \text{and} \quad r > \|F\theta\|/(1-\bar{q}) .$$

5. <u>Evolution equations</u>

The directional contraction method can be applied to prove an existence theorem for nonlinear evolution equations.

Consider the initial value problem

$$\frac{dx}{dt} = F(t,x) , \quad 0 \leq t \leq T , \quad x(0) = \xi , \tag{5.1}$$

where $x = x(t)$ is a function defined on the real interval $[0,T]$ with values in the Banach space X, and $F:[0,T] \times X \to X$. Denote by X_T the space of all continuous functions $x = x(t)$ defined on $[0,T]$ with values in X and with the norm $\|x\|_C = \max[\|x(t)\| : 0 \leq t \leq T]$.

Instead of (5.1), we consider the integral equation

$$x(t) - \int_0^t F(s,x(s))ds = \xi \tag{5.2}$$

as an operator equation in X_T and we assume that the integral operator is closed in X_T.

Suppose that there exists an integrable function $K(t)$, $0 \leq t \leq T$, which has the following property. For arbitrary $x,y \in X_T$ there exists a positive number $\epsilon = \epsilon(x,y) \leq 1$ such that, for all $0 \leq s \leq T$, we have

$$\|F(s,x(s)+\epsilon y(s)) - F(s,x(s))\| \leq K(s)\epsilon\|y\|_C . \tag{5.3}$$

<u>Theorem 5.1</u>. Suppose that the integral operator in (5.2) is closed in X_T and that there exists an integrable function $K(t)$ which satisfies (5.3) and such that

$$\int_0^T K(t)dt = q < 1 . \tag{5.4}$$

Then, for arbitrary $\xi \in X$, equation (5.2) has a continuous solution $x(t)$.

<u>Proof</u>. It is easily seen that condition (5.3) implies condition (4.1) so that the integral operator in (5.2) yields a directional contraction in X_T. Thus, the proof follows immediately from Theorem 4.1.

<u>Definition 5.1</u>. Let $F(t,x)$ be continuous in $(t,x) \in [0,T] \times X$,

and suppose that there exists an integrable function $K(t)$, $0 \leq t \leq T$, which has the following property.

For each $t \in [0,T]$ and $x,y \in X$ there exists a positive $\epsilon = \epsilon(x,y) \leq 1$ such that

$$(5.5) \qquad \|F(t,x+\tau y) - F(t,x)\| < K(t)\tau\|y\|$$

for all $0 \leq \tau \leq \epsilon = \epsilon(x,y) \leq 1$. Then F is said to be Lipschitz demi continuous.

Let us suppose that F satisfies the following hypothesis.

Given two arbitrary continuous functions $x = x(t)$ and $y = y(t) \in X_T$, there exists a sequence $\{\epsilon_n\}$ convergent to zero such that, for any $\eta > 0$, there is a positive δ which satisfies the following condition for almost all n.

$$(5.6) \qquad \epsilon_n^{-1} \int_{\omega} \|F(s,x(s)+\epsilon_n y(s)) - F(s,x(s))\| ds < \eta ,$$

where $\omega \subset [0,T]$ is an arbitrary subset of $[0,T]$ with measure $\mu(\omega) < \delta$.

Theorem 5.2. Suppose that $F:[0,T]\times X \to X$ is Lipschitz demi continuous and satisfies conditions (5.4) and (5.6). Then the initial value problem (5.1) has a solution.

Proof. Let us consider the equivalent problem of solving the integral equation (5.2) for arbitrary $\xi \in X$. The proof will follow from Theorem 4.1 if we can show that the integral operator is a directional contraction. In other words, we have to show that $\epsilon = \epsilon(x,y) \leq 1$ in (5.3) depends on the continuous functions $x,y \in X_T$. Let $\bar{q}$ be arbitrary with $q < \bar{q} < 1$, where q is defined by (5.4). Given arbitrary continuous functions $x = x(t)$, $y = y(t) \in X_T$, put in (5.6) $\eta = (\bar{q}-q)\|y\|_C$ and let δ correspond to η. By a theorem of Lusin, there exists an open subset $\omega \subset [0,T]$ with measure $\mu(\omega) < \delta$ such that the function K is continuous on $[0,T]\setminus\omega$. It follows from (5.5) that for each interior point $s \in [0,T]\setminus\omega$, there exists an open subset $\Delta(s) \subset [0,T]\backslash\omega$ such that

$$\|F(t,x(t)+\tau y(t)) - F(t,x(t))\| < K(t)\|y\|_C$$

for all $t \in \Delta(s)$ and $0 \leq \tau \leq \epsilon = \epsilon(x(s),y(s)) \leq 1$. Let $\{\Delta(s_i)\}$, $i = 1,2,\cdots,n$, be a finite subcovering of $[0,T]\backslash\omega$, and put

$$\epsilon_0 = \epsilon(x,y) = \min[\epsilon(x(s_i),y(s_i)) : 1 \leq i \leq n],$$

where $x,y \in X_T$. Then we have

$$\|F(t,x(t)+\tau y(t)) - F(t,x(t))\| < K(t)\|y\|_C \tag{5.7}$$

for all $t \in [[0,T]\backslash\omega$ and $0 \leq \tau \leq \epsilon_0 \leq 1$. Now, denote by ϵ the number $\epsilon_k \leq \epsilon_0$. Such ϵ_k exists, since $\epsilon_n \to 0$ as $n \to \infty$. Then it follows from (5.6) and (5.7) that

$$\int_0^t \|F(s,x(s)+\epsilon y(s)) - F(s,x(s))\|ds \leq \int_0^T \|F(s,x(s)+\epsilon y(s)) - F(s,x(s))\|ds =$$

$$\int_{[0,T]\backslash\omega} \|F(s,x(s)+\epsilon y(s)) - F(s,x(x))\|ds +$$

$$+ \int_\omega \|F(s,x(s)+\epsilon y(s)) - F(s,x(s))\|ds \leq \epsilon\int_0^T K(s)ds\|y\|_C + (\bar{q}-q)\epsilon\|y\|_C =$$

$$= \bar{q}\epsilon\|y\|_C .$$

Hence, it follows that the integral operator in (5.2) is a directional contraction.

Remark 5.1. If the function $K(t)$ is continuous on $[0,T]$, then condition (5.6) is superfluous.

6. An application of the Krasnosel'skii-Zabreiko theorem to contractor directions

Introduction. Pohožaev [1] has proposed a new solvability principle for nonlinear operator equations. Recently, Browder [1,2] has considerably sharpened and generalized the result of Pohožaev. The result of Pohožaev has also been extended by Zabreiko and Krasnosel'skii [1] by the use of a different method. Browder's method is based on a geometrical theorem and on a concept of asymptotic directions, later generalized by Kirk and Caristi [1] in a combination with a new principle, while Zabreiko and Krasnosel'skii introduced a special equivalent norm. In this section, the concept of contractor directions is used which generalizes Browder's concept of asymptotic directions. Such a generalization makes it possible to enlarge the class of nonlinear operators in question by including operators with directional contractors and contractors with linear majorant functions. The concept of contractor directions is here combined with the method of Zabreiko and Krasnosel'skii. This leads, in addition, to a new proof. All these methods mentioned above are applicable only in the case of an operator with closed range.

Let X be a real vector space, P a mapping of X into the real Banach space Y, x a point in X. Then, by Definition 1.1, the element y of Y is said to belong to the set $\Gamma_x(P)$ of contractor directions for P at x if there exist positive numbers $q < 1$, $\epsilon = \epsilon(x,y) \leq 1$ and an element $h = h(x,y)$ such that

$$\|P(x+h) - Px - \epsilon y\| \leq q\,\epsilon\,\|y\| . \tag{6.1}$$

In our further considerations, the following theorem is essential.

<u>Theorem</u> (Zabreiko-Krasnosel'skii [1]). Let Y be a real Banach space, Z a nonempty proper closed subset of Y and y_0 a point in $Y \setminus Z$. Then there is an equivalent norm $\|\cdot\|_0$ in Y in which the distance from y_0 to Z is achieved (at not more than two points).

Daneš [1] has proved a geometric theorem from which he derives the mentioned theorem of Zabreiko and Krasnosel'skii as well as the geometric theorem of Browder [1]. Daneš also observed that, in the Zabreiko-Krasnosel'skii theorem, the new norm $\|\cdot\|_0$ can be chosen so as to satisfy the following condition,

$$(1-\delta)\|\cdot\| \leq \|\cdot\|_0 \leq \|\cdot\| , \tag{6.2}$$

where $0 < \delta < 1$ is an arbitrary constant.

<u>Lemma 1.1</u>. If $\Gamma_x(P)$ is a set of contractor directions for $P:X \to Y(\|\cdot\|)$ then there exists a positive $\delta < 1$ such that $\Gamma_x(P)$ is a set of contractor directions for P at $x \in X$ in the norm $\|\cdot\|_0$ which satisfies (6.2).

<u>Proof</u>. Let $\Gamma_x(P)$ be the set of contractor directions in the norm $\|\cdot\|$. Then we have, by (6.1) and (6.2),

$$\|P(x+h) - Px - \epsilon y\|_0 \leq \|P(x+h) - Px - \epsilon y\| \leq q\epsilon\|y\| \leq q\epsilon\|y\|_0/(1-\delta) \leq \epsilon\bar{q}\|y\|_0 ,$$

provided $\bar{q}$ is such that $q < \bar{q} < 1$ and δ is chosen so as to satisfy $0 < \delta < (\bar{q}-q)/\bar{q}$. Hence,

$$(6.3) \qquad \|P(x+h) - Px - \epsilon y\|_0 \leq \epsilon\bar{q}\|y\|_0 ,$$

where $h = h(x,y), \epsilon = \epsilon(x,y) \leq 1$ and $0 < \bar{q} < 1$. The proof of the second part of the assertion is similar.

<u>Theorem 6.1</u>. Let X be a real vector space, Y a real Banach space, $P:X \to Y$ a mapping of X into Y such that the range $P(X)$ is closed in Y. Suppose that $y_0 \in Y$ is such that for each x in X, the element $y_0 - Px$ belongs to the closure of the set $\Gamma_x(P)$ of contractor directions for P at x. Then the equation $Px = y_0$ has a solution.

<u>Proof</u>. Let

$$(6.4) \qquad d(y_0, P(X)) > 0 ,$$

where d stands for the distance. By virtue of the Zabreiko-Krasnosel'skii theorem, there exists an equivalent norm $\|\cdot\|_0$ in which the distance a from the point y_0 to the closed set $P(X)$ is achieved at some point Px_0, that is, $a = \|y_0-Px_0\|_0$. By Lemma 6.1, $y_0-Px_0 \in \Gamma_{x_0}(P)$ (in the norm $\|\cdot\|$). The new norm can be chosen so as to satisfy inequalities (6.2) with δ defined as in Lemma 6.1. By Lemma 6.1, y_0-Px_0 belongs to the set $\Gamma_{x_0}(P)$ in the norm $\|\cdot\|_0$.

Let $v = y_0 - Px_0$. Inequality (6.3) with $x = x_0$ and $y = v$ implies there exists a positive number $\epsilon = \epsilon(x_0, v) \leq 1$ and an element $h = h(x_0, v) \in X$ such that

$$\|P(x_0+h) - Px_0 - \epsilon v\|_0 \leq \epsilon\bar{q}\|v\|_0 = \epsilon\bar{q}a < \epsilon a . \tag{6.5}$$

Using inequality (6.5), we obtain

$$a \leq \|P(x_0+h) - y_0\|_0 \leq \|P(x_0+h) - Px_0 - \epsilon v\|_0 + \|y_0 - Px_0 - \epsilon v\|_0 <$$

$$< \epsilon a + (1-\epsilon)\|y_0 - Px_0\|_0 + \epsilon\|y_0 - Px_0 - v\|_0 = a .$$

This is a contradiction and the proof is complete.

Denote by S_r the set $[y: \|y\| < r, y \in Y]$. As a consequence of the above theorem, we obtain

<u>Theorem 6.2</u>. Let X be a real vector space, Y a real Banach space, $P: X \to Y$ a mapping of X into Y such that the range $P(X)$ is closed in Y. Suppose that there exists a finite subset N of X such that for each x in $X \setminus N$, the set $\Gamma_x(P)$ of contractor directions is dense in S_r. Then $P(X) = Y$.

<u>Proof</u>. By Lemma 1.1, $\Gamma_x(P) \supset S_r$ and, consequently, $\Gamma_x(P) = Y$, for all x in $X \setminus N$. Suppose that $P(X) \neq Y$. Then there exists a point y_0 in Y which is not in $P(X)$ such that

$$d(y_0, P(X)) < d(y_0, P(N)) ,$$

since $P(N)$ is finite. Let $\|\cdot\|_0$ be the norm defined in the proof of Theorem 6.1. Then the distance a from the point y_0 to the closed set $P(X)$ is achieved at some point Px_0, that is, by (6.2), $a = \|y_0 - Px_0\|_0 \leq d(y_0, P(X)) < d(y_0, P(N))$, and x_0 is not in N. By Lemma 6.1, $y_0 - Px_0$ belongs to the set $\Gamma_{x_0}(P)$ of contractor directions for $P: X \to Y(\|\cdot\|_0)$. The further reasoning is the same as in the proof of Theorem 6.1.

<u>Definition</u> (Browder [1]). Let X be a real vector space, P a mapping of X into the real Banach space Y, x a point of X. Then the element v of the unit sphere of Y is said to lie in the set $R_x(P)$ of asymptotic directions for P at x if there exists $\xi \neq 0$ in X and a sequence $\{\gamma_j\}$ of positive numbers with $\gamma_j \to 0$ as $j \to \infty$ such that for each j, $P(x+\gamma_j\xi) \neq Px$ and $P(x+\gamma_j\xi) \to Px$ as $j \to \infty$, while $\|P(x+\gamma_j\xi) - Px\|^{-1}(P(x+\gamma_j\xi) - Px) \to v$ as $j \to \infty$.

<u>Theorem</u> (Browder [1]). Let X be a real vector space, Y a real Banach space, P a mapping of X into Y such that $P(X)$ is closed in Y. Suppose that there exists a finite subset N of X such that for all x in $X \backslash N$, the set $R_x(P)$ of asymptotic directions of P at x is dense in the unit sphere of Y. Then $P(X) = Y$.

Since an asymptotic direction of P at $x \in X$ is obviously a contractor direction for P at x, it follows that Browder's theorem is a particular case of Theorem 6.2.

Now let us consider a class of nonlinear mappings $P:X \to Y$ of a real vector space X into a real Banach space Y such that P is differentiable in the sense of Gâteaux, that is, for each $x \in X$ there exists a linear operator $P'(x)$ (the Gâteaux derivative) defined as a limit

$$\lim_{\epsilon \to 0} [P(x+\epsilon h) - Px]/\epsilon = P'(x)h \quad \text{for any} \quad h \in X . \tag{6.6}$$

<u>Lemma 6.2</u>. Let X be a real vector space, Y a real Banach space, and let P be a mapping of X into Y which is differentiable in the Gâteaux sense. Suppose that for $x \in X$, the range $P'(x)(X)$ is dense in Y. Then $\Gamma_x(P) = Y$.

<u>Proof</u>. Let $C < 1$ be a positive constant. Then for every y of Y, there exists an element $h = h(x,y) \in X$ such that

$$\|P'(x)h - y\| \leq C\|y\| , \quad 0 < C < 1 . \tag{6.7}$$

Choose q with $C < q < 1$ and a positive η such that $\eta \leq q-C$. For arbitrary $y \in Y$ there is an element $h \in X$ satisfying (6.7). By (6.6), there exists a positive number $\epsilon \leq 1$ such that

$$\|P(x+\epsilon h) - Px - \epsilon P'(x)h\| \leq \epsilon\eta\|y\| .$$

Then we obtain

$$\|P(x+\epsilon h) - Px - \epsilon y\| \leq \|P(x+\epsilon h) - Px - \epsilon P'(x)h\| + \epsilon\|P'(x)h - y\| \leq$$

$$\leq \epsilon\eta\|y\| + \epsilon C\|y\| \leq q\epsilon\|y\| ,$$

by virtue of (6.7). Thus, y is a contractor direction for P at $x \in X$. Since y is an arbitrary element of Y, it follows that $\Gamma_x(P) = Y$.

Remark 6.1. Let us observe that condition (6.7) is equivalent to the fact that the range $P'(x)(X)$ is dense in Y. For by using (6.7), we can define an infinite sequence $\{x_n\}$ of elements of X such that

$$\|P'(x)x_{n+1}-y\| \leq C\|P'(x)x_n-y\| .$$

In fact, by (6.7), there exists an element $h_{n+1} \in X$ such that

$$\|P'(x)h_{n+1} - [P'(x)x_n-y]\| \leq C\|P'(x)x_n-y\| .$$

Since

$$\|P'(x)h_{n+1} - [P'(x)x_n-y]\| = \|P'(x)(x_n-h_{n+1})-y\| ,$$

we can put $x_{n+1} = x_n-h_{n+1}$. Putting $x_1 = h$, where h is chosen so as to satisfy inequality (6.7), for the arbitrary but fixed $y \in Y$. Then we obtain

$$\|P'(x)x_n-y\| \leq C^n\|y\|$$

for $n = 1,2,\cdots$.

Obviously, $P'(x)$ in Remark 6.1 can be replaced by an arbitrary linear bounded operator satisfying (6.7).

The following theorem is a particular case of Theorem 6.2.

Theorem (Zabreiko-Krasnosel'skii [1]). Let X be a real vector space, Y a real Banach space, $P:X \to Y$ a mapping of X into Y such that the range $P(X)$ is closed in Y. If P is differentiable in the Gâteaux sense at each $x \in X$ and the range $P'(x)(X)$ is dense in Y, then $P(X) = Y$.

Proof. It is evident that for each $x \in X$, $P'(x)$ satisfies the hypotheses of Lemma 6.2. Hence, the proof follows from Theorem 6.2.

Let X be a vector space, Y a Banach space and P a mapping of X into Y. Then a linear mapping $\Gamma(x):Y \to X$ is said to be a directional contractor for P at x if there exists a positive $q < 1$ such that for every $y \in Y$ there exists a positive number $\epsilon = \epsilon(x,y) \leq 1$ such that

$$\|P(x+\epsilon\Gamma(x)y) - Px - \epsilon y\| \leq q\,\epsilon\,\|y\| .$$

If for all $x \in X$, $y \in Y$ (see Chapter 4).

Theorem 6.3. Let X be a real vector space, Y a real Banach space, P a mapping of X into Y such that $P(X)$ is closed in Y. If there exists a directional contractor of P at each $x \in X$ (except for a finite subset of X), then $P(X) = Y$.

Proof. The proof follows from Theorem 6.2, since it is obvious that $\Gamma_x(P) = Y$, by virtue of Definition 2.1, Chapter 4.

Theorem 6.4. Let X, Y be real Banach spaces, $P:X \to Y$ a continuous mapping such that $P(X)$ is closed in Y. Suppose that for each $x \in X$, $\Gamma(x):Y \to X$ is a continuous linear mapping of Y onto X satisfying the following inequality,

$$\|P(x+\Gamma(x)y) - Px - y\| \leq q\|y\| \quad \text{for all } y \in Y \text{ with } 0 < q < 1 .$$

Then P is a homeomorphism of X onto Y.

Proof. It follows from Theorem 6.2 that P maps X onto Y. The remaining assertions result from the inequality

$$(1-q)\|y\| \leq \|P(x+\Gamma(x)y) - Px\| \quad \text{for all} \quad x \in X , \ y \in Y .$$

Remark 6.2. Since $P(X)$ is closed, in Theorems 6.3 and 6.4, the condition

$$\|\Gamma(x)\| \leq B \quad \text{for all} \quad x \in X ,$$

where B is some constant, is not required.

CHAPTER 6

CLASSIFIED CONTRACTOR DIRECTIONS

Introduction. The general method of contractor directions seems to be a natural generalization of the method of directional contractors. However, in order to make the method applicable to nonlinear differential and integral equations, a further development of the concept of contractor directions is necessary. Therefore, certain conditions are imposed on the sets of contractor directions. In this way, a special class of increasing continuous functions is involved in the definition of specialized contractor directions. This class is closely connected with the classical Cauchy integral test for infinite series. By using the method of specialized contractor directions sufficient conditions are obtained for general existence theorems of solutions of nonlinear equations. This method does not require the operator to have closed range. However, if the range of the operator in question is closed, then no additional conditions are imposed on the sets of contractor directions. In this particular case, the method is much simpler. Thus, the unified theory also yields a generalization of results obtained by Pohožaev [1], Browder [1], Zabreiko and Krasnosel'skii [1], Kirk and Caristi [1]. All the methods used by the authors mentioned above are different and applicable only in the case of an operator with closed range.

1. A general solvability principle

Let $P:D(P) \subset X \to Y$ be a nonlinear mapping, where $D(P)$ is a vector space and X, Y are real or complex Banach spaces. Denote by $\mathbb{B}$ the class of increasing continuous functions B such that

(i) $\qquad B(0) = 0 \,, \quad B(s) > 0 \quad \text{for} \quad s > 0 \,;$

(ii) $\int_0^a s^{-1}B(s)\,ds < \infty$ for some $a > 0$;

and

(iii) $0 < \gamma < 1$ implies $B(e^{-\gamma t})t \to 0$ as $t \to \infty$.

Let g be a continuous function such that $g(t) > 0$ for $t > 0$.

Definition 1.1. $\Gamma_x(P) = \Gamma_x(P,q)$ is a set of contractor directions at $x \in D(P)$ for $P: D(P) \subset X \to Y$, which has the (B,g)-property, if for arbitrary $y \in \Gamma_x(P)$ there exist a positive number $\epsilon = \epsilon(x,y) \leq 1$ and an element $h \in X$ such that

$$\|P(x+\epsilon h) - Px - \epsilon y\| \leq q\epsilon\|y\| \tag{1.1}$$

$$\|h\| \leq B(g(\|x\|)\cdot\|y\|) , \tag{1.2}$$

where $x+\epsilon h \in D(P)$ and $q = q(P) < 1$ is some positive constant independent of $x \in D(P)$.

Definition 1.2. The nonlinear mapping $P: D(P) \subset X \to Y$ is (B,g)-differentiable at $x \in D(P)$ if there is a dense subset $V \subset Y$ such that for arbitrary $y \in V$ there exists an element $h \in X$ which satisfies condition (1.2) and $\|P(x+\epsilon h) - Px - \epsilon y\|/\epsilon \to 0$ as $\epsilon \to 0+$.

Given an element $x_0 \in D(P)$, a continuous function g such that $g(t) > 0$ for $t > 0$, and a function $B \in \mathbb{B}$, let us put

$$\bar{r} = 2(1-q)^{-1}\int_0^b s^{-1}B(s)\,ds , \quad b = e^{1-q} ,$$

where $0 < q < 1$ is the same as in Definition 1.1. We can find a polynomial $\bar{p}$ such that

$$|g(t) - \bar{p}(t)| \leq 1 \quad \text{for} \quad 0 \leq t \leq \|x_0\| + \bar{r} .$$

Then we have

(iv) $g(t) \leq p(t) = \sum_{i=0}^{N} c_i t^i$ for $0 \leq t \leq \|x_0\| + \bar{r}$, where p dominates $\bar{p} + 1$ and has nonnegative coefficients. Now put

$$W(\|x_0\|) = \frac{1}{2}\sum_{i=0}^{N} c_i \sum_{k=0}^{i} \binom{i}{k} \|x_0\|^{2k} \tag{v}$$

(vi) $$C = \sum_{i=0}^{N} 2^i c_i$$

(vii) $$\rho = 1/2N \quad \text{and} \quad \beta = 1/2C .$$

Let $M \geq 1$ be such that

(viii) $$B(e^{W(\|x_0\|)+1}\|Px_0\|e^{-\frac{1}{2}(1-q)t})t < \rho\beta(1-q) \quad \text{for} \quad t \geq M .$$

A local existence theorem

Let $S = S(x_0, r)$ be an open ball with center $x_0 \in X$ and radius r, and put $U = D(P) \cap \overline{S}$, where $\overline{S}$ is the closure of S.

Theorem 1.1. Suppose that the following hypotheses are satisfied:

(1.3) $$P: U \to Y \text{ is closed on } U ;$$

(1.4) for each $x \in U_0 = D(P) \cap S$, a set $\Gamma_x(P)$ of contractor directions with the (B,g)-property exists, being dense in some ball with center 0 in Y;

(1.5) $$r \geq 2(1-q)^{-1}\int_0^a s^{-1}B(s)\,ds , \quad a = e^{1-q}e^{W(\|x_0\|)}\|Px_0\| ;$$

(1.6) $$B(e^{W(\|x_0\|)+1}\|Px_0\|)M < \rho\beta(1-q) \quad \text{and} \quad e^{W(\|x_0\|)}\|Px_0\| < 1 ,$$

where $W(\|x_0\|)$, M and ρ, β are defined by (i) - (viii), respectively. Then the equation $Px = 0$ has a solution $x \in U$.

Proof. The proof is based upon a further development of the transfinite induction argument.

We construct well-ordered sequences of positive numbers t_α and elements $x_\alpha \in D(P)$ as follows. Put $t_0 = 0$, and let x_0 be the given element. Suppose that t_γ and x_γ have been constructed for all $\gamma < \alpha$, provided, for arbitrary ordinal numbers $\gamma < \alpha$, inequalities

(1.7_γ) $$\|Px_\gamma\| \leq e^{-(1-q)t_\gamma}\|Px_0\| ,$$

(1.8_γ) $$\|x_\gamma\| \leq \|x_0\| + \beta(1-q)t_\gamma^\rho ,$$

are satisfied, where $\rho = 1$ if $t_\gamma < 1$, and $0 < \rho < 1$ if $t_\gamma \geq 1$,

$0 < \beta < 1$. The constants ρ and β will be determined below.

For first kind ordinal numbers, $\beta = \gamma + 1 < \alpha$, the following inequalities are satisfied:

$$(1.9_{\gamma+1}) \qquad 0 < t_{\gamma+1} - t_\gamma \leq 1$$

$$(1.10_{\gamma+1}) \qquad \|x_{\gamma+1} - x_\gamma\| \leq (t_{\gamma+1} - t_\gamma) B(e^{W(\|x_0\|)} \|Px_0\| e^{-\frac{1}{2}(1-q)t_\gamma}) ,$$

and

$$(1.11_{\gamma+1}) \qquad \|Px_{\gamma+1} - Px_\gamma\| \leq (1+q) \|Px_0\| e^{-(1-q)t_\gamma} (t_{\gamma+1} - t_\gamma) ;$$

and, for second kind (limit) ordinal numbers, $\gamma < \alpha$, the following relations hold:

$$(1.12_\gamma) \qquad t_\gamma = \lim_{\beta \nearrow \gamma} t_\beta , \quad x_\gamma = \lim_{\beta \nearrow \gamma} x_\beta , \quad Px_\gamma = \lim_{\beta \nearrow \gamma} Px_\beta .$$

Then it follows from (1.10), (1.12), Lemmas 3.1 and 3.2, Chapter 4, that, for arbitrary $\lambda < \gamma < \alpha$, we have

$$\|x_\gamma - x_\lambda\| \leq \sum_{\lambda \leq \beta < \gamma} \|x_{\beta+1} - x_\beta\| \leq \sum_{\lambda \leq \beta < \gamma} (t_{\beta+1} - t_\beta) B(e^{W(\|x_0\|)} \|Px_0\| e^{-\frac{1}{2}(1-q)t_\beta}) =$$

$$= \sum_{\lambda \leq \beta < \gamma} (t_{\beta+1} - t_\beta) B(e^{W(\|x_0\|)} \|Px_0\| e^{\frac{1}{2}(1-q)(t_{\beta+1} - t_\beta)} e^{-\frac{1}{2}(1-q)t_{\beta+1}}) <$$

$$< \sum_{\lambda \leq \beta < \gamma} (t_{\beta+1} - t_\beta) B(e^{1-q} e^{W(\|x_0\|)} \|Px_0\| e^{-\frac{1}{2}(1-q)t_{\beta+1}}) \leq$$

$$\leq \sum_{\lambda \leq \beta < \gamma} \int_{t_\beta}^{t_{\beta+1}} B(e^{1-q} e^{W(\|x_0\|)} \|Px_0\| e^{-\frac{1}{2}(1-q)t}) dt =$$

$$= \int_{t_\lambda}^{t_\gamma} B(e^{1-q} e^{W(\|x_0\|)} \|Px_0\| e^{-\frac{1}{2}(1-q)t}) dt .$$

Hence, we obtain the following estimate

$$(1.13) \qquad \|x_\gamma - x_\lambda\| \leq \int_{t_\lambda}^{t_\gamma} B(e^{1-q} . e^{W(\|x_0\|)} \|Px_0\| e^{-\frac{1}{2}(1-q)t}) dt .$$

In the same way, we obtain from (1.9), (1.11), Lemmas 3.1 and 3.2, Chapter 4, that

$$\|Px_\gamma - Px_\lambda\| \le (1+q)e^{1-q}\|Px_0\|\int_{t_\lambda}^{t_\gamma} e^{-(1-q)t}dt \,. \tag{1.14}$$

Suppose that α is a first kind ordinal number. If $Px_{\alpha-1} = 0$, then the proof of the theorem is completed.

If $Px_{\alpha-1} \neq 0$, then we put

$$t_\alpha = t_{\alpha-1} + \tau_\alpha \,, \tag{1.15}$$

and

$$x_\alpha = x_{\alpha-1} + \tau_\alpha h_\alpha \,, \tag{1.16}$$

where h_α and $\tau_\alpha = \epsilon \le 1$ are chosen so as to satisfy (1.1) and (1.2) with $x = x_{\alpha-1}$, $h = h_\alpha$ and $y = -Px_{\alpha-1}$. Hence, we obtain, by virtue of $(1.7_{\alpha-1})$ and (1.1) with $x = x_{\alpha-1}$, $y = -Px_{\alpha-1}$, $h = h_\alpha$,

$$\|Px_\alpha\| \le (1-\tau_\alpha)\|Px_{\alpha-1}\| + q\tau_\alpha\|Px_{\alpha-1}\| =$$

$$= (1-(1-q)\tau_\alpha)\|Px_{\alpha-1}\| < e^{-(1-q)\tau_\alpha}\|Px_{\alpha-1}\| \le e^{-(1-q)t_\alpha}\|Px_0\| \,,$$

i.e.,

$$\|Px_\alpha\| \le e^{-(1-q)t_\alpha}\|Px_0\| \,. \tag{1.17_α}$$

Now we consider two cases; case a), where $t_\alpha < 1$ and case b), where $t_\alpha \ge 1$. In both cases we have

$$p(\|x_\alpha\|) = \sum_{i=0}^{N} c_i\|x_\alpha\|^i \le \sum_{i=0}^{N} c_i(\|x_0\| + \beta(1-q)t_\alpha^\rho)^i =$$

$$= \sum_{i=0}^{N} c_i \sum_{k=0}^{i} \binom{i}{k}\|x_0\|^k[\beta(1-q)t_\alpha^\rho]^{i-k} \le \frac{1}{2}\sum_{i=0}^{N} c_i \sum_{k=0}^{i}\binom{i}{k}\|x_0\|^{2k} +$$

$$+ \frac{1}{2}\sum_{i=0}^{N} c_i \sum_{k=0}^{i}\binom{i}{k}[\beta(1-q)t_\alpha^\rho]^{2(i-k)} \le W(\|x_0\|) + \frac{1}{2}\beta(1-q)Ct_\alpha \,.$$

Hence, we obtain

$$p(\|x_\alpha\|) \le W(\|x_0\|) + \frac{1}{2}(1-q)t_\alpha \tag{1.18_α}$$

provided (1.8_α) holds true, where

$$W(\|x_0\|) = \frac{1}{2}\sum_{i=0}^{N} c_i \sum_{k=0}^{i} \binom{i}{k} \|x_0\|^{2k} \tag{1.19}$$

$$C = \sum_{i=0}^{N} c_i \sum_{k=0}^{i} \binom{i}{k} = \sum_{i=0}^{N} 2^i c_i , \tag{1.20}$$

and where ρ and β are chosen so as to satisfy

$$\rho 2N = 1 \quad \text{and} \quad \beta C = 1/2 . \tag{1.21}$$

In case a), we have, by (1.15), (1.16), (1.2), $(1.8_{\alpha-1})$ and (iv) - (viii),

$$\|x_\alpha\| \leq \|x_{\alpha-1}\| + \tau_\alpha B(g(\|x_{\alpha-1}\|)\cdot\|y_{\alpha-1}\|) \leq$$

$$\leq \|x_0\| + \beta(1-q)t_{\alpha-1}^{\rho} + \tau_\alpha B(e^{W(\|x_0\|)+\frac{1}{2}(1-q)t_{\alpha-1}}\|Px_0\|e^{-(1-q)t_{\alpha-1}}) .$$

Hence,

$$\|x_\alpha\| \leq \|x_0\| + \beta(1-q)t_{\alpha-1}^{\rho} + \tau_\alpha B(e^{W(\|x_0\|)}\|Px_0\|e^{-\frac{1}{2}(1-q)t_{\alpha-1}}) , \tag{1.22}$$

and

$$\|x_\alpha\| \leq \|x_0\| + \beta(1-q)t_{\alpha-1}^{\rho} + \tau_\alpha B(e^{W(\|x_0\|)}\|Px_0\|) . \tag{1.23}$$

Hence, it follows that

$$\|x_\alpha\| \leq \|x_0\| + \beta(1-q)(t_{\alpha-1}+\tau_\alpha)$$

or condition (1.8_α) will be satisfied with $\rho = 1$ if

$$B(e^{W(\|x_0\|)}\|Px_0\|) \leq \beta(1-q) \tag{1.24}$$

is satisfied, where β is defined by (1.21) and (1.20).

Now let us consider case b). It is easy to see that (1.8_α) will be satisfied if

$$\beta(1-q)t_{\alpha-1}^{\rho} + \tau_\alpha B(e^{W(\|x_0\|)}\|Px_0\|e^{-\frac{1}{2}(1-q)t_{\alpha-1}}) \leq \beta(1-q)(t_{\alpha-1}+\tau_\alpha)^{\rho}$$

or

$$(1.25) \qquad B(e^{W(\|x_0\|)}\|Px_0\|e^{-\frac{1}{2}(1-q)t_{\alpha-1}}) < \rho\beta(1-q)(t_{\alpha-1}+\tau_\alpha)^{\rho-1} ,$$

by virtue of (1.22). Therefore, replacing $t_{\alpha-1}+\tau_\alpha$ in (1.25) by t, let $M \geq 1$ be such that

$$(1.26) \qquad B(e^{W(\|x_0\|)+1}\|Px_0\|e^{-\frac{1}{2}(1-q)t})t < \rho\beta(1-q) \quad \text{for} \quad t \geq M ,$$

by virtue of (viii), and let

$$(1.27) \qquad B(e^{W(\|x_0\|)+1}\|Px_0\|)M < \rho\beta(1-q) ,$$

where ρ and β are defined by (1.20), (1.21). Then, obviously (1.25) will be satisfied, and consequently, (1.8_α) will be satisfied, too. Thus, in both cases, conditions (1.26) and (1.27) imply that (1.8_α) will be satisfied, where α is an ordinal number of first kind.

It follows from (1.15), (1.16), (1.2), $(1.7_{\alpha-1})$, $(1.18_{\alpha-1})$ and (iv) that

$$(1.28_\alpha) \qquad \|x_\alpha - x_{\alpha-1}\| \leq (t_\alpha - t_{\alpha-1})B(e^{W(\|x_0\|)}\|Px_0\|e^{-\frac{1}{2}(1-q)t_{\alpha-1}}) .$$

Thus, we have shown that if α is a first kind ordinal number, then the induction assumptions hold true for α. Now suppose that α is an ordinal number of second kind and put $t_\alpha = \lim_{\gamma\nearrow\alpha} t_\gamma$. Let $\{\gamma_n\}$ be an increasing sequence convergent to α. It follows from (1.13) and (1.14) that $\{x_{\gamma_n}\}$ and $\{Px_{\gamma_n}\}$ are Cauchy sequences and so are $\{x_\gamma\}$ and $\{Px_\gamma\}$. Denote by x_α and y_α their limits, respectively. Since P is closed on U, we infer that $x_\alpha \in U$ and $y_\alpha = Px_\alpha$, provided $x_\gamma \in U$. If $t_\alpha < \infty$, then the limit passage in (1.7_{γ_n}) and (1.8_{γ_n}) yields (1.7_α) and (1.8_α), respectively. The relationships (1.12_α) are satisfied by definition of t_α and x_α, since $y_\alpha = Px_\alpha$. This process will terminate if $t_\alpha = \infty$, where α is of second kind. In this case, $Px_\alpha = 0$, by virtue of (1.7_α). The limit x_α exists, by (1.13), since

$$(1.29) \qquad \int_0^\infty B(e^{1-q}e^{W(\|x_0\|)}\|Px_0\|e^{-\frac{1}{2}(1-q)t})dt = 2(1-q)^{-1}\int_0^a s^{-1}B(s)ds < \infty ,$$

where $a = e^{1-q} e^{W(\|x_0\|)} \|Px_0\|$. Finally, it follows from (1.13), (1.29) and (1.5) that all $x_\gamma \in U_0$. It results from (1.5) and (1.6) that

$$(1.30) \qquad 2(1-q)^{-1} \int_0^a s^{-1} B(s)\,ds \le 2(1-q)^{-1} \int_0^{e^{1-q}} s^{-1} B(s)\,ds \ .$$

Hence, it follows that $g(\|x_\alpha\|) \le p(\|x_\alpha\|)$, i.e., condition (iv) is valid for $t = \|x_\alpha\|$, and as a consequence, we also obtain that (1.18_α) is true. This completes the proof.

Consider now a particular case of Theorem 1.1.

Theorem 1.1a. Suppose that the hypotheses (1.3) - (1.5) are satisfied, where $B(s) = s$, $g(t) = C(t+1)$ for some constant $C > 0$ and

$$(1.5a) \qquad r \ge 2C(\|x_0\|+1)\|Px_0\| e^{(1-q)/2}/(1-q) \ ,$$

where $\|x_0\|$ should be replaced by 1 if $\|x_0\| \le 1$.

$$(1.6a) \qquad C\|Px_0\| \le (1-q)/4 \ .$$

Then, equation $Px = 0$ has a solution $x \in U$.

Proof. The general method of proof is similar to that of Theorem 1.1. However, we have to replace the induction assumption (1.8_γ) by the following one.

$$(1.8'_\gamma) \qquad \|x_\gamma\| \le \|x_0\| e^{\beta(1-q)t_\gamma} \ , \quad \beta = 1/2 \ ,$$

where $\|x_0\|$ should be replaced by 1 if $\|x_0\| \le 1$. Then we obtain the following estimate,

$$\|x_\gamma - x_\lambda\| \le \sum_{\lambda \le \beta < \gamma} \|x_{\beta+1} - x_\beta\| \le \sum_{\lambda \le \beta < \gamma} (t_{\beta+1} - t_\beta) C(\|x_\beta\|+1)\|Px_\beta\| \le$$

$$\le \sum_{\lambda \le \beta < \gamma} (t_{\beta+1} - t_\beta) C(\|x_0\| e^{\beta(1-q)t_\beta} + 1)\|Px_0\| e^{-(1-q)t_\beta} =$$

$$= \sum_{\lambda \le \beta < \gamma} (t_{\beta+1} - t_\beta) C(\|x_0\|\|Px_0\| e^{-\frac{1}{2}(1-q)t_\beta} + \|Px_0\| e^{-(1-q)t_\beta}) \le$$

$$\leq \sum_{\lambda\leq\beta<\gamma} (t_{\beta+1}-t_\beta)C(\|x_0\|+1)\|Px_0\|e^{-\frac{1}{2}(1-q)t_\beta} \leq$$

$$\leq \sum_{\lambda\leq\beta<\gamma} (t_{\beta+1}-t_\beta)C(\|x_0\|+1)\|Px_0\|e^{(1-q)/2}e^{-\frac{1}{2}(1-q)t_{\beta+1}} \leq$$

$$\leq \sum_{\lambda\leq\beta<\gamma} C(\|x_0\|+1)\|Px_0\|e^{(1+q)/2}\int_{t_\beta}^{t_{\beta+1}} e^{-\frac{1}{2}(1-q)t}\,dt .$$

Hence, we obtain

$$\|x_\gamma - x_\lambda\| \leq Ce^{(1-q)/2}(\|x_0\|+1)\|Px_0\|\int_{t_\lambda}^{t_\gamma} e^{-\frac{1}{2}(1-q)t}\,dt \leq$$

$$\leq 2Ce^{(1-q)/2}(\|x_0\|+1)\|Px_0\|/(1-q) \leq r .$$

It remains to show the induction passage for $(1.8'_\gamma)$. Suppose that α is an ordinal number of first kind. We have in case of $\|x_0\| > 1$,

$$\|x_\alpha\| \leq \|x_{\alpha-1}\| + \tau_\alpha C(\|x_{\alpha-1}\|+1)\|Px_{\alpha-1}\| \leq$$

$$\leq \|x_{\alpha-1}\| + \tau_\alpha C(\|x_0\|e^{\beta(1-q)t_{\alpha-1}}+1)\|Px_0\|e^{-(1-q)t_{\alpha-1}} \leq$$

$$\leq \|x_{\alpha-1}\| + \tau_\alpha 2C\|x_0\|\cdot\|Px_0\|e^{-\frac{1}{2}(1-q)t_{\alpha-1}} .$$

Hence, it follows that

$$\|x_\alpha\| \leq \|x_0\|e^{\beta(1-q)t_\alpha}, \quad \beta = 1/2$$

if

$$\|x_0\|e^{\beta(1-q)t_{\alpha-1}} + \tau_\alpha 2C\|x_0\|\cdot\|Px_0\|e^{-\frac{1}{2}(1-q)t_{\alpha-1}} \leq \|x_0\|e^{\beta(1-q)t_\alpha} ,$$

where $t_\alpha = t_{\alpha-1} + \tau_\alpha$. Consider the function

$$\varphi(\tau) = e^{\beta(1-q)(t+\tau)} - \tau 2C\|Px_0\|e^{-\frac{1}{2}(1-q)t} - e^{\beta(1-q)t} .$$

The derivative $\varphi'(\tau) > 0$ if $C\|Px_0\| < (1-q)/4$. Hence, it follows that (1.8_α) holds true. Consider now the case, where $\|x_0\| \leq 1$. Then we have

$$\|x_\alpha\| \leq \|x_{\alpha-1}\| + \tau_\alpha C(\|x_{\alpha-1}\|+1)\|Px_{\alpha-1}\| \leq$$

$$\leq e^{\beta(1-q)t_{\alpha-1}} + \tau_\alpha C(e^{\beta(1-q)t_{\alpha-1}}+1)\|Px_0\|e^{-(1-q)t_{\alpha-1}} \leq$$

$$\leq e^{\beta(1-q)t_{\alpha-1}} + \tau_\alpha 2C\|Px_0\|e^{-(1-q)t_{\alpha-1}} .$$

Thus, we have to show that

$$e^{\beta(1-q)t_{\alpha-1}} + \tau_\alpha 2C\|Px_0\|e^{-\frac{1}{2}(1-q)t_{\alpha-1}} \leq e^{\beta(1-q)(t_{\alpha-1}+\tau_\alpha)} .$$

But this inequality is exactly the same as in the case where $\|x_0\| > 1$. Hence, it follows that condition (1.8_α) holds true in both cases. The further reasoning is the same as in the proof of Theorem 1.1.

Remark 1.1. In Theorem 1.1, if $B(s) = s$ for $s \geq 0$, then the first inequality in (1.6) can be replaced by the following one.

$$e^{W(\|x_0\|)}\|Px_0\| < \rho\beta(1-q)^2/2e^{(1-q)/2} , \tag{1.31}$$

where $W(\|x_0\|)$, ρ, β are defined by (1.19), (1.20), (1.21), respectively.

Proof. Put $a = e^{W(\|x_0\|)}\|Px_0\|$, then, by virtue of (1.22), we have to prove that

$$\beta(1-q)t_{\alpha-1}^{\rho} + \tau_\alpha ae^{-\frac{1}{2}(1-q)t_{\alpha-1}} \leq \beta(1-q)(t_{\alpha-1}+\tau_\alpha)^{\rho} . \tag{1.32}$$

Let us consider case b), where $0 < \rho < 1$, $t \geq 1$, since case a) yields the same condition (1.24). Consider the function

$$\varphi(\tau) = \beta(1-q)(t+\tau)^{\rho} - \tau ae^{-\frac{1}{2}(1-q)t} - \beta(1-q)t^{\rho} ,$$

which satisfies the condition $\varphi(0) = 0$, and its derivative is positive if

$$\varphi'(\tau) = \rho\beta(1-q)(t+\tau)^{\rho-1} - ae^{-\frac{1}{2}(1-q)t} > 0 ,$$

that is, if

$$a < \rho\beta(1-q)e^{\frac{1}{2}(1-q)t}(t+\tau)^{\rho-1} , \quad 0 < \tau \leq 1 , \quad t \geq 1 . \tag{1.33}$$

But it is easy to see that

$$p\beta(1-q)^2/2e^{(1-q)/2} < \rho\beta(1-q)e^{\frac{1}{2}(1-q)t}(t+\tau)^{\rho-1} .$$

Thus, it follows from the last inequality that if (1.31) is satisfied, then so is (1.33) and, consequently, condition (1.32) holds true.

Remark 1.2. Condition (1.4) can be replaced by the requirement that P is (B,g)-differentiable at every $x \in U_0$.

Proof. It follows from Definitions 1.1 and 1.2 that, for every $x \in U_0$, P has a set $\Gamma_x(P)$ of contractor directions with the (B,g)-property and $\Gamma_x(P)$ is dense in Y .

Theorem 1.2. Suppose that the hypotheses (1.3), (1.4) and (1.5), where $a = e^{1-q}$, and, in addition, $Px_0 = y_0$. Then there exists a ball K with center y_0 such that for every $y \in K$, the equation $Px = y$ has a solution $x \in U$.

Proof. Denote by $\overline{P}$ the operator with values $\overline{P}x = Px - y$. Then there exists a ball K with center y_0 such that condition (1.6) will be satisfied for $\overline{P}$ if $y \in K$. Thus, all hypotheses of Theorem 1.1 are satisfied.

A global existence theorem

We now assume that $P:D(P) \subset X \to Y$ satisfies the following condition.

(α) If the sequence $\{x_n\} \subset D(P)$ is not bounded, then $\{Px_n\}$ contains no Cauchy sequence.

A mapping which satisfies condition (α) will be briefly called a Cauchy mapping.

Theorem 1.3. Suppose that the following hypotheses are satisfied.

1) The graph of the nonlinear mapping $P:D(P) \subset X \to Y$ is closed in $X \times Y$.

2) For each $x \in D(P)$, a set $\Gamma_x(P)$ of contractor directions

with the (B,g)-property exists, which is dense in some ball with center 0 in Y.

3) P is a Cauchy mapping.

Then P is a mapping onto Y.

Proof. If the range $P(D(P))$ is closed, then a more general theorem is true (see Chapter 5). Thus there exists an element y_0 that is not in $P(D(P))$ and a sequence $\{x_n\} \subset D(P)$ such that

$$\|Px_n - y_0\| \to 0 \quad \text{as} \quad n \to \infty \tag{1.34}$$

$$\|x_n\| \leq c \quad \text{for} \quad n = 1,2,\cdots , \tag{1.35}$$

where c is some constant, by virtue of condition 3). Consider the operator $\overline{P}$ with values $\overline{P}x = Px - y_0$. With $\overline{r}$ defined by condition (iv), we approximate $g(t)$ on the closed interval $0 \leq t \leq c + \overline{r}$ by the polynomial $\overline{p}(t)$ and define $p(t)$ as in (iv). Then we put $\overline{x}_0 = x_m$, where x_m is an element of the sequence $\{x_n\}$ which satisfies inequalities (1.6) with x_0 replaced by x_m. Such an element exists for $\overline{P}$, since $\overline{P}x_n \to 0$ as $n \to \infty$. With such a choice of $\overline{x}_0$ for $\overline{P}$, all hypotheses of Theorem 1.1 are satisfied and the equation $\overline{P}x = Px - y_0 = 0$ has a solution. Therefore, our assumption that y_0 is not in $P(D(P))$ leads to a contradiction which proves the theorem.

As a consequence of Theorem 1.3, we obtain the following.

Theorem 1.4. A closed Cauchy mapping $P: D(P) \subset X \to Y$ which is (B,g)-differentiable is a mapping onto Y.

Proof. The proof follows from Remark 1.2.

As a consequence of Theorem 1.1a, we obtain the following two theorems which do not require P to be a Cauchy mapping.

Theorem 1.5. Suppose that conditions 1) and 2) of Theorem 1.3 are satisfied, where $B(s) = s$ and $g(t) = C(t+1)$ with some constant $C > 0$. Then $P: D(P) \subset X \to Y$ is a mapping onto Y.

Proof. The proof follows from that of Theorem 1.3 and from Theorem 1.1a.

Theorem 1.6. A closed mapping $P:D(P) \subset X \to Y$ which is (B,g)-differentiable, where

$$B(s) = s \quad \text{and} \quad g(t) = C(t+1)$$

for some constant $C > 0$, is a mapping onto Y.

Proof. The proof follows from that of Theorem 1.6, from Theorem 1.1a and Remark 1.2.

2. Non-bounded directional contractors

The global existence theorems proved in Section 1 can be applied to nonlinear operators having directional contractors which may not be bounded. For the definition of a directional contractor, see Chapter 4. Denote by $L(Y \to X)$ the set of all linear continuous mappings from the Banach space Y into the Banach space X.

Theorem 2.1. A nonlinear closed Cauchy mapping $P:D(P) \subset X \to Y$ which has a directional contractor $\Gamma:D(P) \to L(Y \to X)$ such that

$$\|\Gamma(x)\| \leq g(\|x\|) \quad \text{for all} \quad x \in D(P) , \tag{2.1}$$

where g is some continuous function, is a mapping onto Y.

Proof. The proof follows from Theorem 1.3, where $B(s) = s$.

We consider now nonlinear operators which are differentiable in the Gâteaux sense.

Theorem 2.2. Let $P:D(P) \subset X \to Y$ be a nonlinear closed Cauchy mapping. Suppose that for each $x \in D(P)$, the Gâteaux derivative $P'(x)$ is an additive and homogeneous operator which has a continuous inverse $\Gamma(x) = P'(x)^{-1}$ such that

$$\|\Gamma(x)\| \leq g(\|x\|) \quad \text{for all} \quad x \in D(P) ,$$

where g is some continuous function. Then P is a mapping onto Y.

Proof. The proof follows from Theorem 2.1, since $\Gamma:D(P) \to L(Y \to X)$ is a directional contractor for P and satisfies condition (2.1).

Theorem 2.3. Let $P:D(P) \subset X \to Y$ be a nonlinear closed Cauchy mapping which is differentiable in the sense of Fréchet. Suppose that for each $x \in D(P)$, the Fréchet derivative $P'(x)$ is a mapping onto Y. If there exists a continuous function g such that

$$\|[P'(x)^*]^{-1}\| \leq g(\|x\|) \quad \text{for all} \quad x \in D(P) ,$$

where $*$ indicates the adjoint, then P is a mapping onto Y.

Proof. The proof follows from Theorem 1.4 and Lemma 2.1, Chapter 5, since P is (B,g)-differentiable with $B(s) = s$.

Remark 2.1. In Theorems 2.1 - 2.3, the condition that P is a Cauchy mapping can be omitted if $g(t) = C(t+1)$ for some constant $C > 0$.

Theorem 2.4. Let $P:D(P) \subset X \to Y$ be a nonlinear closed Cauchy mapping which is differentiable in the Fréchet sense with Hölder continuous derivative $P'(x)$, i.e., there exist positive numbers K, $\alpha \leq 1$ such that

$$\|P'(x) - P'(\bar{x})\| \leq K\|x-\bar{x}\|^{\alpha} \quad \text{for all} \quad x,\bar{x} \in D(P) . \tag{2.2}$$

Moreover, for every $x \in D(P)$, let $A(x):X \to Y$ be a bounded linear nonsingular operator such that

$$\|A(x)^{-1}\| \leq g(\|x\|) \quad \text{and} \quad \|P'(x) - A(x)\| \leq c(\|x\|) , \tag{2.3}$$

for all $x \in D(P)$, where g and c are some functions, g being continuous. Suppose that there exist positive constants r and $q < 1$ such that

$$(1+\alpha)^{-1}K[g(\|x\|)]^{1+\alpha}r + c(\|x\|)g(\|x\|) \leq q < 1 \quad \text{for all} \quad x \in D(P) . \tag{2.4}$$

Then P is a mapping onto Y.

Proof. The proof follows from Theorem 2.1, since $\Gamma(x) = A(x)^{-1}$ is a directional contractor satisfying condition (2.1). In fact, we have, by (2.2) - (2.4),

$$\|P(x+\Gamma(x)y) - Px - y\| \leq \|P(x+\Gamma(x)y) - Px - P'(x)\Gamma(x)y\| +$$

$$+ \|P'(x)\Gamma(x)y - A(x)\Gamma(x)y\| \leq (1+\alpha)^{-1}K\|\Gamma(x)y\|^{1+\alpha} +$$

$$+ c(\|x\|)\|\Gamma(x)y\| \leq (1+\alpha)^{-1}K[g(\|x\|)]^{1+\alpha}\|y\|^{1+\alpha} + c(\|x\|)g(\|x\|)\|y\| \leq$$

$$\leq q\|y\| \quad \text{if} \quad \|y\| \leq r^{1/\alpha} .$$

Theorem 2.5. Let $P:D(P) \subset X \to Y$ be a nonlinear closed Cauchy mapping and let $T:D(P) \subset X \to Y$ be an operator differentiable in the Fréchet sense with Hölder continuous Fréchet derivative $T'(x)$, i.e., there exist positive constants K and $\alpha \leq 1$ such that

$$\|T'(x) - T'(\bar{x})\| \leq K\|x-\bar{x}\|^{\alpha}, \quad 0 < \alpha \leq 1, \tag{2.5}$$

for all $x \in D(P)$. Moreover, suppose that $T'(x)$ is nonsingular and that there exist a continuous function g and a function c such that

$$\|T'(x)^{-1}\| \leq g(\|x\|) \quad \text{and} \quad \|(Px-Tx) - (P\bar{x}-T\bar{x})\| \leq c(\|x\|)\|x-\bar{x}\|, \tag{2.6}$$

for all $x,\bar{x} \in D(P)$. If there exists a positive constant $q < 1$ such that condition (2.4) is satisfied, then P is a mapping onto Y.

Proof. The proof follows from Theorem 2.1, since $\Gamma(x) = T(x)^{-1}$ is a directional contractor satisfying condition (2.1). In fact, we have, by (2.5), (2.6) and (2.4),

$$\|P(x+\Gamma(x)y) - Px - y\| \leq \|T(x+\Gamma(x)y) - Tx - T'(x)\Gamma(x)y\| +$$

$$+ \|[P(x+\Gamma(x)y) - T(x+\Gamma(x)y)] - [Px-Tx]\| \leq$$

$$\leq (1+\alpha)^{-1}K\|\Gamma(x)y\|^{1+\alpha} + c(\|x\|)\|\Gamma(x)y\| \leq$$

$$\leq (1+\alpha)^{-1}K[g(\|x\|)]^{1+\alpha}\|y\|^{1+\alpha} + c(\|x\|)g(\|x\|)\|y\| \leq$$

$$\leq [(1+\alpha)^{-1}K[g(\|x\|)]^{1+\alpha}r + c(\|x\|)g(\|x\|)]\|y\| \leq q\|y\|$$

if $\|y\| \leq r^{1/\alpha}$.

Theorem 2.6. Let $P:D(P) \subset X \to Y$ be closed and differentiable in the sense of Fréchet with Hölder continuous Frechet derivative $P'(x)$ satisfying (2.2). Suppose that for each $x \in D(P)$, there exists a bounded linear operator $\Gamma(x):Y \to X$ such that

(2.7) $$\|P'(x)\Gamma(x)-I\| \leq C\|Px\| \quad \text{for all} \quad x \in D(P)$$

and

(2.8) $$\|\Gamma(x)\| \leq B \quad \text{for all} \quad x \in D(P) ,$$

where I is the identity mapping of Y and C , B are some constants. Then P is a mapping onto Y .

Proof. The proof follows from Theorem 2.1 since $\Gamma:D(P) \to L(Y \to X)$ is a bounded directional contractor for P . In fact, we have, by (2.2), (2.7) and (2.8),

$$\|P(x+\Gamma(x)y) - Px - y\| \leq \|P(x+\Gamma(x)y) - Px - P'(x)\Gamma(x)y\| +$$

$$+ \|P'(x)\Gamma(x)y - y\| \leq [(1+\alpha)^{-1}KB^{1+\alpha}\|y\|^{\alpha} + C\|y\|]\|y\| \leq$$

$$\leq [(1+\alpha)^{-1}KB^{1+\alpha}r^{\alpha} + Cr]\|y\| \leq q\|y\| \quad \text{if} \quad \|y\| \leq r \leq 1$$

and $(1+\alpha)^{-1}KB^{1+\alpha}(1+C)r^{\alpha} \leq q$ with $0 < q < 1$. This is true for all y replaced by $-Px$. But this restriction is admissible.

Let us notice that a local existence theorem can also be obtained by means of condition (2.7) which is less restrictive than the existence of a right inverse of $P'(x)$.

The following remark is evident.

Remark 2.2. Remark 2.1 applies also to Theorems 2.4 and 2.5.

3. The method of contractor directions in a scale of Banach spaces

3.1. Let us consider a family of Banach spaces $V^\rho (0 \leq \rho \leq p)$ in the order $V^0 \supset V^\rho \supset V^p$, satisfying the condition

$$\|v\|_\rho \leq c\|v\|_0^{1-\rho/p}\|v\|_p^{\rho/p} \tag{3.1}$$

where c depends on ρ, p only and $\|\cdot\|_\rho$ is the norm in V^ρ. For examples and applications of such families of spaces, see Moser [1]. Let Y be a Banach space (real or complex) with norm $\|\cdot\|$.

Definition 3.1. A mapping

$$P:D(P) \subset V^\rho(\ 0 \leq \rho \leq p) \rightarrow Y$$

is said to be approximately differentiable at $x \in D(P)$ in the directional sense or briefly A-differentiable if there exists a positive μ which has the following property. For arbitrary $y \in Y$, and for arbitrary $K > 1$, $Q > 1$, there exist elements $z \in Y$ and $h \in V^p$ such that

$$\|P(x+\epsilon h) - Px - \epsilon z\|/\epsilon \rightarrow 0 \quad \text{as} \quad \epsilon \rightarrow 0+ , \tag{3.2}$$

$$\|z-y\| \leq KQ^{-\mu}\|y\| , \tag{3.3}$$

$$\|h\|_p \leq \overline{C}KQ\|y\| , \tag{3.4}$$

$$\|h\|_0 \leq \overline{C}\|z\| , \tag{3.5}$$

where $x + \epsilon h \in D(P)$ and $\overline{C}$ is some constant. A similar case of A-differentiability is discussed by Moser [1], i.e., when P has a derivative P' defined by

$$P'(x)h = \lim_{t\rightarrow 0} t^{-1}[P(x+th) - Px] \tag{*}$$

which admits approximate solutions satisfying conditions (3.3) - (3.5), where $z = P'(x)h$.

Lemma 3.1. Suppose that

$$P: D(P) \subset V^{\rho} (0 \le \rho \le p) \to Y$$

is A-differentiable at all $x \in D(P)$. Then, for arbitrary positive $q < 1$, $\Gamma_x(P) = \Gamma_x(P,q) = Y$ for all $x \in D(P)$, where P is considered as a mapping $P: D(P) \subset V^{\rho} \to Y$, ρ being arbitrary with $0 \le \rho < p$. The function $B(s)$ which is involved in the definition of the set $\Gamma_x(P)$ of special contractor directions is of the form

$$B(s) = Cs^{\sigma} \quad \text{with} \quad 0 < \sigma < 1 ,$$

where C is some positive constant.

Proof. By Lemma 1.1, Chapter 5, it is sufficient to show that $\Gamma_x(P)$ contains a ball $S(0,r) \subset Y$ with radius $r < 1$. Let $x \in D(P) \subset V^{\rho}$ be fixed along with $0 < q < 1$. For $y \in S(0,r)$, put

$$K = \|y\|^{-\beta} \quad \text{and} \quad Q = (q/2)^{-1/\mu} \|y\|^{-\beta/\mu} . \tag{3.6}$$

Then

$$KQ^{-\mu} = q/2 \quad \text{and} \quad KQ = (q/2)^{-1/\mu} \|y\|^{-\beta(1+1/\mu)} . \tag{3.7}$$

By virtue of Definition 3.1, there exist $z \in Y$ and $h \in V^p$ which satisfy the relations (3.2) - (3.5). Put $\bar{x} = x + \epsilon h$. Then we have

$$\|P(x+\epsilon h) - Px - \epsilon y\| \le \|P(x+\epsilon h) - Px - \epsilon z\| +$$

$$+ \epsilon\|z-y\| \le q\epsilon\|y\|/2 + q\epsilon\|y\|/2 = q\epsilon\|y\| ,$$

by (3.6), (3.7) and (3.2) if $\epsilon > 0$ is sufficiently small. On the other hand, we have, by virtue of (3.1) - (3.7),

$$\|h\|_{\rho} \le c\|h\|_0^{1-\rho/p} \cdot \|h\|_p^{\rho/p} \le C\|y\|^{\sigma} , \tag{3.8}$$

where $C = c\bar{C}(1+q/2)^{1-\rho/p}(q/2)^{-\rho/\mu p}$,

$$(3.9) \qquad \sigma = 1 - \beta\rho/p - \beta\rho/\mu p \quad \text{and} \quad 0 < \sigma < 1 \quad \text{if}$$

$$(3.10) \qquad 0 < \beta < (p/\rho)\mu/(1+\mu) \ .$$

Hence, it follows that if β in (3.6) is chosen so as to satisfy (3.10), then the function $B(s)$ can be defined by the formula $B(s) = Cs^{\sigma}$ with $0 < \sigma < 1$ defined by (3.9) and the constant C determined in formula (3.8). It is obvious that $B \in \mathbb{B}$, i.e., $B(s) > 0$ for $s > 0$ is a continuous increasing function which satisfies conditions (ii) and (iii). The case $\rho = 0$ is evidently contained in the discussion of Section 1.

Theorem 3.1. Let $P:D(P) \subset V^{\rho}(0 \leq \rho \leq p) \to Y$ be a mapping which is A-differentiable at all $x \in D(P)$. If $0 \leq \rho < p$ is such that $P:D(P) \subset V^{\rho} \to Y$ is a closed mapping, then for arbitrary $y \in Y$, the equation $Px = y$ has a solution $x \in V^{\rho}$.

Proof. The proof follows immediately from Lemma 3.1 and Theorem 1.2, where the space X is replaced by the Banach space V^{ρ} .

3.2. In order to apply the general method of contractor directions to some solutions of partial differential equations, the following change in the Definition 3.1 will be needed.

Definition 3.2. A mapping

$$P:D(P) \subset V^{\rho}(0 \leq \rho \leq p) \to G^{\sigma}(0 \leq \sigma \leq s)$$

is A-differentiable at $x \in D(P)$ if there exists a positive μ which has the following property. For arbitrary $y \in G^{\sigma}(0 \leq \sigma \leq s)$ and for arbitrary $K > 1$, $Q > 1$, there exist elements $z \in G^{0}$ and $h \in V^{p}$ such that

$$(3.11) \qquad \|P(x+\epsilon h) - Px - \epsilon z\|_{0}/\epsilon \to 0 \quad \text{as} \quad \epsilon \to 0+$$

(3.12) $$\|z-y\|_0 \leq KQ^{-\mu}\|y\|_0$$

(3.13) $$\|h\|_p \leq KQ\|y\|_0$$

(3.14) $$\|h\|_0 \leq \overline{C}\|z\|_0$$

provided

(3.15) $$\|y\|_\sigma \leq K \quad \text{and} \quad \|x\|_p < K ,$$

where $x + \epsilon h \in D(P)$ and $\overline{C}$ is some positive constant.

This definition of A-differentiability is clearly more general than the Definition 3.1, and it generalizes the case discussed by Moser [1], where P has a derivative P' which is defined by formula (*) and which admits approximate solutions h satisfying conditions (3.12) - (3.15) with $z = P'(x)h$. It is assumed that both families $V^\rho (0 \leq \rho \leq p)$ and $G^\sigma (0 \leq \sigma \leq s)$ of Banach spaces satisfy the same conditions as in Section 3.1. The following is a local existence theorem for A-differentiable mappings.

<u>Theorem 3.2</u>. Given a mapping

$$P: D(P) \subset V^\rho (0 \leq \rho \leq p) \to G^\sigma (0 \leq \sigma \leq s) ,$$

let $S = S(x_0, r) \subset V^\rho$ with some $\rho < p$ such that there exist positive numbers q, β, $M > 1$ with $q < 1$, $\beta \leq \mu$ and

(3.16) $$\|Px_0\|_0 < \beta(1-q)(q/2)^{-1/\mu} M^{-(1+\mu)/\mu} \|x_0\|_p^{-1/\mu} ;$$

(3.17) $$\|Px\|_\sigma \leq MK \quad \text{if} \quad \|x\|_p < K , \quad x \in S .$$

The mapping P is A-differentiable at every $x \in S \subset V^\rho$, where

(3.18) $$r \geq C\|Px_0\|_0 e^{(1-\delta)(1-q)}/(1-\delta)(1-q) ,$$

where

$$\delta = (\rho/p)\beta(1+\mu)/\mu < 1$$

and

$$C = [\overline{C}(1+q/2)]^{1-\rho/p}[M\|x_0\|_p]^{(1+\mu)/\mu}(2/q)^{1/\mu} ;$$

(if $\beta = \mu$, then $\rho < p(1+\mu)$) . The graph of P is closed in V $V^\rho \times G^0$ and $V^p \times G^0$. Then the equation $Px = 0$ has a solution $x \in S \subset V^\rho$. If $\beta < \mu/(1+\mu)$, then $x \in V^p$.

Proof. The general pattern of proof of Theorem 1.1 is adopted here. Thus, we construct well-ordered sequences of positive numbers t_α and elements $x_\alpha \in S$ as follows. Put $t_0 = 0$ and suppose that t_γ and x_γ have been constructed for all $\gamma < \alpha$, provided, for arbitrary ordinal numbers, the inequalities

$$(3.19_\gamma) \qquad \|Px_\gamma\|_0 \le e^{-(1-q)t_\gamma}\|Px_0\|_0$$

$$(3.20_\gamma) \qquad \|x_\gamma\|_p \le \|x_0\|_p e^{\beta(1-q)t_\gamma} , \quad t_\gamma < \infty ,$$

are satisfied and for first kind ordinal numbers, $\beta = \gamma + 1 < \alpha$, the following inequalities are satisfied:

$$(3.21_{\gamma+1}) \qquad 0 < \tau_{\gamma+1} = t_{\gamma+1} - t_\gamma \le 1 ,$$

$$(3.22_{\gamma+1}) \qquad \|Px_{\gamma+1} - Px_\gamma\|_0 \le (1+q)\|Px_0\| e^{-(1-q)t_\gamma}(t_{\gamma+1} - t_\gamma) ;$$

$$(3.23_{\gamma+1}) \qquad \|x_{\gamma+1} - x_\gamma\|_p \le C_1(1+q/2)\|Px_0\|_0 e^{-(1-q)t_\gamma}(t_{\gamma+1} - t_\gamma) ,$$

where $C_1 = M^{(1+\mu)/\mu}(2/q)^{1/\mu}\|x_0\|_p^{(1+\mu)/\mu}$;

$$(3.24_{\gamma+1}) \qquad \|x_{\gamma+1} - x_\gamma\|_0 \le \overline{C}(1+q/2)\|Px_0\|_0 e^{-(1-q)t_\gamma}(t_{\gamma+1} - t_\gamma) ,$$

where the constant $\overline{C}$ is the same as in (3.14), and for second kind ordinal numbers, $\gamma < \alpha$, the following relationships hold:

$$(3.25_\gamma) \qquad \begin{gathered} t_\gamma = \lim_{\beta \nearrow \gamma} t_\beta , \quad t_\gamma < \infty , \quad x_\gamma = \lim_{\beta \nearrow \gamma} x_\beta \quad (\text{in } V^p) , \\ Px_\gamma = \lim_{\beta \nearrow \gamma} Px_\beta \quad (\text{in } G^0) . \end{gathered}$$

Then it follows from (3.23), (3.25), Lemmas 3.1 and 3.2, Chapter 4, that, for arbitrary $\lambda < \gamma < \alpha$ such that $t_\alpha < \infty$, we have

$$\|x_\gamma - x_\lambda\|_p \leq \sum_{\lambda\leq\beta<\gamma} \|x_{\beta+1} - x_\beta\|_p \leq c_1 \|Px_0\|_0 \sum_{\lambda\leq\beta<\gamma} \int_{t_\beta}^{t_{\beta+1}} e^{-[1-\beta(1+\mu)/\mu](1-q)t} dt$$

$$= c_1 \|Px_0\|_0 \int_{t_\lambda}^{t_\gamma} e^{-[1-\beta(1+\mu](1-q)t} dt \; .$$

Hence, we obtain the following estimate:

$$\|x_\gamma - x_\lambda\|_p \leq c_1 \|Px_0\|_0 \int_{t_\lambda}^{t_\gamma} e^{-[1-\beta(1+\mu)/\mu](1-q)t} dt \; . \tag{3.26}$$

In the same way, we obtain from (3.21), (3.22), Lemmas 3.1 and 3.2, Chapter 4, that

$$\|Px_\gamma - Px_\lambda\|_0 \leq (1+q)e^{1-q} \|Px_0\|_0 \int_{t_\lambda}^{t_\gamma} e^{-(1-q)t} dt \; . \tag{3.27}$$

Now we obtain from (3.24), (3.23) and (3.1) the following estimate:

$$\|x_{\gamma+1} - x_\gamma\|_\rho \leq c_2 \|Px_0\|_0 e^{-(1-\delta)(1-q)t_\gamma} (t_{\gamma+1} - t_\gamma) =$$

$$= c_2 \|Px_0\|_0 e^{(1-\delta)(1-q)(t_{\gamma+1}-t_\gamma)} e^{-(1-\delta)(1-q)t_{\gamma+1}} (t_{\gamma+1} - t_\gamma) <$$

$$< c_2 \|Px_0\|_0 e^{(1-\delta)(1-q)} \int_{t_\gamma}^{t_{\gamma+1}} e^{-(1-\delta)(1-q)t} dt \; ,$$

where $\delta = (\rho/p)\beta(1+\mu)/\mu < 1$, by (3.18), and $c_2 = c[\overline{c}(1+q/2)]^{1-\rho/p} c_1^{\rho/p}$. Hence, we have

$$\|x_\gamma - x_\lambda\|_\rho \leq c_2 \|Px_0\|_0 e^{(1-\delta)(1-q)} \int_{t_\lambda}^{t_\gamma} e^{-(1-\delta)(1-q)t} dt \; . \tag{3.28}$$

It follows in particular from (3.28) that

$$\|x_\gamma - x_0\|_\rho \leq c_2 \|Px_0\|_0 e^{(1-\delta)(1-q)} \int_0^\infty e^{-(1-\delta)(1-q)t} dt =$$

$$= c_2 \|Px_0\|_0 e^{(1-\delta)(1-q)} / (1-\delta)(1-q) \leq r \; , \tag{3.29_γ}$$

by (3.18).

Suppose that α is a first kind ordinal number. If $Px_{\alpha-1} = 0$, then the proof of the theorem is completed. If $Px_{\alpha-1} \neq 0$, then we put

$$t_\alpha = t_{\alpha-1} + \tau_\alpha , \quad x_\alpha = x_{\alpha-1} + \tau_\alpha h , \tag{3.30}$$

where $h \in V^p$ and $\tau_\alpha = \epsilon \leq 1$ along with $z \in G^0$ are chosen so as to satisfy (3.13) and (3.14) with $y = -Px_{\alpha-1}$, $K = MK_{\alpha-1} = $ $= M\|x_0\|_p e^{\beta(1-q)t_{\alpha-1}}$ (see (3.20_γ)) and Q such that $MK_{\alpha-1}Q^{-\mu} = q/2$ or $Q = (MK_{\alpha-1}2/q)^{1/\mu}$. Then we have

$$\|P(x_{\alpha-1} + \epsilon h) - Px_{\alpha-1} - \epsilon y\|_0 \leq \|P(x_{\alpha-1} + \epsilon h) - Px_{\alpha-1} - \epsilon z\|_0 +$$

$$+ \epsilon\|z-y\|_0 \leq \frac{1}{2} q\epsilon\|y\|_0 + \epsilon MK_{\alpha-1}Q^{-\mu}\|y\|_0 = q\epsilon\|y\|_0 ,$$

for sufficiently small $\epsilon = \tau_\alpha \leq 1$, by virtue of (3.11), that is,

$$\|P(x_{\alpha-1} + \tau_\alpha h) - Px_{\alpha-1} - \tau_\alpha y\| \leq q\tau_\alpha\|y\|_0 , \tag{3.31}$$

for $y = -Px_{\alpha-1}$, since then condition (3.15) is satisfied with $x = x_{\alpha-1}$, $y = -Px_{\alpha-1}$ and $K = MK_{\alpha-1}$, by virtue of (3.17), $(3.20_{\alpha-1})$ and $(3.29_{\alpha-1})$. Hence, we obtain, by (3.31), $(3.19_{\alpha-1})$ and (3.30)

$$\|Px_\alpha\|_0 \leq (1-\tau_\alpha)\|Px_{\alpha-1}\|_0 + q\tau_\alpha\|Px_{\alpha-1}\|_0 =$$

$$= (1-(1-q)\tau_\alpha)\|Px_{\alpha-1}\|_0 < e^{-(1-q)\tau_\alpha}\|Px_{\alpha-1}\|_0 \leq$$

$$\leq e^{-(1-q)t_\alpha}\|Px_0\|_0 ,$$

i.e.,

$$\|Px_\alpha\|_0 \leq \|Px_0\|_0 e^{-(1-q)t_\alpha} . \tag{3.32_α}$$

Further, it follows from (3.30), $(3.19_{\alpha-1})$ and (3.31) with $y = -Px_{\alpha-1}$ that

$$\|Px_\alpha - Px_{\alpha-1}\|_0 \leq (1+q)\tau_\alpha\|Px_{\alpha-1}\|_0 \leq (1+q)\|Px_0\|_0 e^{-(1-q)t_{\alpha-1}}(t_\alpha - t_{\alpha-1}) .$$

Thus, condition (3.22_α) is satisfied. It remains to show that the induction assumptions (3.23_α), (3.24_α) and (3.20_α) are also satisfied.

By virtue of (3.30), (3.32_α), (3.14) and (3.12) with $y = -Px_{\alpha-1}$, $K = MK_{\alpha-1}$, we have

$$\|x_\alpha - x_{\alpha-1}\|_0 = \tau_\alpha \|h\|_0 \le \tau_\alpha \bar{c}(1+MK_{\alpha-1}Q^{-\mu})\|Px_{\alpha-1}\|_0 \le$$

$$\le \bar{c}(1+q/2)\|Px_0\|_0 e^{-(1-q)t_{\alpha-1}}(t_\alpha - t_{\alpha-1}) .$$

It follows from (3.30), (3.23_α) and (3.13) with $y = -Px_{\alpha-1}$, $K = MK_{\alpha-1}$ and $Q = (MK_{\alpha-1}2/q)^{1/\mu}$ that

$$(3.33) \qquad \|x_\alpha - x_{\alpha-1}\|_p = \tau_\alpha \|h\|_p \le \tau_\alpha (MK_{\alpha-1})^{(1+\mu)/\mu}(2/q)^{1/\mu}\|Px_{\alpha-1}\|_0 \le$$

$$\le C_1\|Px_0\|_0 e^{-[1-\beta(1+\mu)/\mu](1-q)t_{\alpha-1}}(t_\alpha - t_{\alpha-1}) ,$$

where C_1 is the same as in ($3.23_{\gamma+1}$). Now, by virtue of (3.30), ($3.20_{\alpha-1}$) and (3.33) we have

$$(3.33a) \qquad \|x_\alpha\|_p \le \|x_{\alpha-1}\|_p + \tau_\alpha\|h\|_p \le \|x_0\|_p e^{\beta(1-q)t_{\alpha-1}} +$$

$$+ \tau_\alpha C_1 \|Px_0\|_0 e^{-[1-\beta(1+\mu)/\mu](1-q)t_{\alpha-1}} =$$

$$= (1 + \tau_\alpha C_1 \|x_0\|_p^{-1}\|Px_0\|_0 e^{-(1-\beta/\mu)(1-q)t_{\alpha-1}})\|x_0\|_p e^{\beta(1-q)t_{\alpha-1}} \le$$

$$\le (1+\tau_\alpha C_1\|x_0\|_p^{-1}\|Px_0\|_0)\|x_0\|_p e^{\beta(1-q)t_{\alpha-1}} \le \|x_0\|_p e^{\beta(1-q)t_\alpha} \quad \text{if}$$

$$1 + \tau_\alpha C_1 \|x_0\|_p^{-1}\|Px_0\|_0 \le e^{\beta(1-q)\tau_\alpha} .$$

The last inequality is satisfied if

$$\beta(1-q) > C_1\|x_0\|_p^{-1}\|Px_0\|_0$$

or

$$\beta(1-q) > M^{(1+\mu)/\mu}\|x_0\|_p^{1/\mu}(\tfrac{2}{q})^{1/\mu}\|Px_0\|_0 ,$$

since then we obtain that the derivative of the function

$\varphi(\tau) = e^{\beta(1-q)\tau} - \tau C_1\|x_0\|_p^{-1}\|Px_0\|_0 - 1$ is positive. Hence, we conclude that if condition (3.16) is satisfied, then inequality (3.20_γ) holds true for $\gamma = \alpha$. Now, we suppose that α is an ordinal number of second kind and put $t_\alpha = \lim_{\gamma \nearrow \alpha} t_\gamma$, where $t_\alpha < \infty$. Let $\{\gamma_n\}$ be an increasing sequence convergent to α . It follows from (3.26) and (3.27) that $\{x_{\gamma_n}\}$ and $\{Px_{\gamma_n}\}$ are Cauchy sequences and so are $\{x_\gamma\}$ and $\{Px_\gamma\}$. Denote by x_α and y_α their limits in V^p , G^0 , respectively. Since the graph of P is closed in $V^p \times G^0$, we infer that $x_\alpha \in D(P)$, $x_\alpha \in V^p$ and $y_\alpha = Px_\alpha$. If $t_\alpha < \infty$, then the limit passage in (3.19_{γ_n}) and (3.20_{γ_n}) yields (3.19_α) and (3.20_α) , respectively. The relationships (3.25_α) are satisfied by the definition of t_α and x_α , since $y_\alpha = Px_\alpha$. This process will terminate if $t_\alpha = \infty$, where α is of second kind. In this case $\{x_{\gamma_n}\}$ is a Cauchy sequence in V^ρ , by virtue of the estimate (3.28), since the integral $\int_0^\infty e^{-(1-\delta)(1-q)t}dt$ is convergent. Denote by x_α the limit in V^ρ of $\{x_{\gamma_n}\}$. It follows from (3.29) that $x_\alpha \in S(x_0,r) \subset V^\rho$, where $\rho < p$ is specified in (3.18). By virtue of (3.27), $\{Px_{\gamma_n}\}$ is a Cauchy sequence in G^0 and has a limit $y \in G^0$. Since the graph of P is closed in $V^\rho \times G^0$ by assumption, it follows that $y = Px_\alpha$. On the other hand, we have that $Px_\alpha = 0$, by virtue of (3.19). Therefore, the proof of the theorem is completed.

Remark 3.1. The inequality (3.16) is clearly satisfied in some neighborhood of Px_0 in G^0 and, consequently, there exists a solution of Px_0 in G^0 and, consequently, there exists a solution for every y in that neighborhood of the equation $Px = y$. Then Px_0 has to be replaced by Px_0-y and the proof remains the same.

Remark 3.2. For the sake of simplicity, we assumed that the domain $D(P)$ of P contained a ball $S = S(x_0,r) \subset V^\rho$. As in the case of local existence theorems, we can consider P defined on $U = D(P) \cap \overline{S}$

and A-differentiable at each $x \in U_0 = D(P) \cap S$.

Remark 3.3. Suppose that P has a derivative P' defined by

$$P'(x)h = \lim_{t \to 0} t^{-1}[P(x+th) - Px]$$

which admits approximate solutions h which satisfy conditions (3.12) - (3.15) with $z = P'(x)h$, then P is clearly A-differentiable.

Moser [1] has proved a local existence theorem for a mapping P which possesses a derivative operator $P'(x)$ admitting approximate solutions h in the above sense. His method which is an iterative process is different and requires $P'(x)$ to satisfy the additional condition

$$\|P(x+h) - Px - P'(x)h\|_0 \leq M\|h\|_0^{2-\beta} \cdot \|h\|_p^{\beta} , \tag{3.34}$$

where $\|x-x_0\|_0 < 1$, $h \in V^p$, if $\|h\|_0 < \|h\|_p$, and β is such that $0 < \beta < 1$ and satisfies certain inequality.

Various applications and examples of approximate solutions in the above sense also are given by Moser [1]. In particular, he applied there his general local existence theorem to nonlinear systems of partial differential equations which are a generalization of linear positive symmetric systems. To avoid the difficulties of boundary behavior, he discusses there differential equations on the torus. The systems under his consideration are of the form

$$F_k(x,u,u_x) = 0 \quad \text{for} \quad k = 1,2,\cdots,m , \tag{3.35}$$

where F_k are of neriod 2π in $x_1,x_2,\cdots,x_n$ and admit sufficiently many derivatives in $|u| + |p| \leq 1$, where p has nm many components p_k corresponding to $\partial u_k/\partial x$.

He introduces matrices

$$a_{k\ell}^{(\nu)}(x,y,p) = \partial F_k/\partial p_{\ell\nu} = \partial F_\ell/\partial p_{k\nu}$$

for $k,\ell = 1,\cdots,m$; $\nu = 1,\cdots,n$, and

$$b_{k\ell}(x,y,p) = \partial F_k/\partial u_\ell ,$$

where he requires that the $a_{k\ell}^{(\nu)}$ are symmetric matrices.

For such systems, Moser [1] proves an existence theorem which is based on his general local existence theorem mentioned above. Under proper hypotheses, he proves the existence of approximate solutions for the derivative operator of the system (3.35). Condition (3.34) is also satisfied for this system.

It follows from Moser's discussion that a similar existence theorem can also be obtained for the system (3.35) by means of Theorem 3.2. In this case, there is no need to require condition (3.34) to be satisfied.

Another application of Moser's method was found by Rabinowitz [1]. He established the existence of periodic solutions for the nonlinear equation

$$u_{tt} - u_{xx} + 2u_t + \epsilon F(x,t,u_t,u_{xx},u_{xt},u_{tt}) = 0$$

with boundary conditions

$$u(x,t+\tau) = u(x,t) , \quad u(0,t) = u(1,t) = 0 , \quad 0 \leq x \leq 1$$

provided F is sufficiently smooth, periodic in t, with period τ, and ϵ is sufficiently small.

It follows from Rabinowitz's investigation that the same result can also be obtained by means of Theorem 3.2.

3.3. We can replace the family $G^\sigma (0 \leq \sigma \leq s)$ of Banach spaces by one Banach space Y. Then, in the Definition 3.2, the inequality $\|y\|_\sigma \leq K$ in (3.15) has to be dropped.

Theorem 3.2 remains valid if $G^\sigma (0 \leq \sigma \leq s)$ is replaced by a Banach space Y and assumption (3.17) is no longer required.

The proof in this case is without change.

4. Another variant of the method of contractor directions in a scale of Banach spaces

4.1. By introducing the concept of α-differentiability, another variant of the method described in Section 3, can be proposed.

Definition 4.1. A mapping

$$P:D(P) \subset V^{\rho}(0 \leq \rho \leq p) \to G^{\sigma}(0 \leq \sigma \leq s)$$

is α-differentiable at $x \in D(P)$ if there exists a positive Q which has the following property. For arbitrary $y \in G^{\sigma}(0 \leq \sigma \leq s)$ and for arbitrary K, there exists an element $h \in V^{p}$ such that

$$\|P(x+\epsilon h) - Px - y\|_0/\epsilon \to 0 \quad \text{as} \quad \epsilon \to 0+ \tag{4.1}$$

$$\|h\|_p \leq QK\|y\|_0 \tag{4.2}$$

$$\|h\|_0 \leq \overline{C}\|y\|_0 \tag{4.3}$$

provided

$$\|y\|_\sigma \leq K \quad \text{and} \quad \|x\|_p < K \, , \tag{4.4}$$

where $x + \epsilon h \in D(P)$ and $\overline{C}$ is some positive constant.

Suppose that P has a derivative operator P' which is defined by formula (*). Then we say that P' has the α-property if there exists a constant Q such that for arbitrary $y \in G^{\sigma}(0 \leq \sigma \leq s)$ and for arbitrary K, the equation $P'(x)h = y$ has a solution $h \in V^{p}$ which satisfies conditions (4.2) - (4.4). It is assumed that both families $V^{\rho}(0 \leq \rho \leq p)$ and $G^{\sigma}(0 \leq \sigma \leq s)$ of Banach spaces satisfy the same conditions as in Section 3.1. Clearly if P has a derivative $P'(x)$ with the α-property, then P is α-differentiable at $x \in D(P)$. The following is a local existence theorem for α-differentiable mappings.

Theorem 4.1. Given a mapping

$$P:D(P) \subset V^{\rho}(0 \leq \rho \leq p) \to G^{\sigma}(0 \leq \sigma \leq s) \, ,$$

let $S = S(x_0, r) \subset V^\rho$ for some $\rho < p$ be such that there exist positive numbers $M > 1$, $q < 1$, $\beta < 1$ with

$$\|Px_0\|_0 < \beta(1-q)/QM ; \tag{4.5}$$

$$\|Px\|_\sigma \le M \quad \text{if} \quad \|x\|_p < K , \quad x \in S . \tag{4.6}$$

The mapping P is α-differentiable at every $x \in S \subset V^\rho$, where

$$r \ge C_2 \|Px_0\|_0 e^{(1-\delta)(1-q)}/(1-\delta)(1-q) , \quad \delta = \beta\rho/p < 1 ,$$
$$C_2 = \overline{C}^{1-\rho} \cdot C_1^{\rho/p} , \quad C_1 = M\|x_0\|_p . \tag{4.8}$$

The graph of P is closed in $V^\rho \times G$ and $V^p \times G^0$. Then the equation $Px = 0$ has a solution $x \in S \subset V^\rho$.

Proof. The proof is only slightly different from that of Theorem 3.2. Thus, proceeding as in the proof of Theorem 3.2 we replace there the induction assumption $(3.22_{\gamma+1})$ by the following one:

$$\|x_{\gamma+1} - x_\gamma\|_p \le C_1 \|Px_0\|_0 e^{-(1-\beta)(1-q)t_\gamma} (t_{\gamma+1} - t_\gamma) , \tag{4.9}$$

where $C_1 = M\|x_0\|_p$. We also replace $(3.23_{\gamma+1})$ by the following inequality:

$$\|x_{\gamma+1} - x_\gamma\|_0 \le \overline{C} \|Px_0\|_0 e^{-(1-q)t_\gamma} (t_{\gamma+1} - t_\gamma) . \tag{4.10}$$

These changes lead to new estimates replacing (3.26) and (3.28) by the following inequalities, respectively:

$$\|x_\gamma - x_\lambda\|_p \le C_1 \|Px_0\|_0 \int_{t_\lambda}^{t_\gamma} e^{-(1-\beta)(1-q)t} dt \tag{4.11}$$

$$\|x_\gamma - x_\lambda\|_\rho \le C_2 \|Px_0\|_0 \int_{t_\lambda}^{t_\gamma} e^{-(1-\delta)(1-q)t} dt , \tag{4.12}$$

where δ and C_2 are defined in (4.8). Thus, the estimate (3.29_γ) is valid with δ and C_2 defined in (4.8). Furthermore, inequality (3.31) follows from (4.1) for arbitrary $0 < q < 1$ if ϵ is sufficiently small. Replacing the corresponding induction assumptions by (4.9), (4.10) and using the estimates (4.11), (4.12), we continue the proof in the same way as that of Theorem 3.2.

Let us observe that Remark 3.1 with (3.16) replaced by (4.5) is also valid here as well as Remark 3.2. With regard to Remark 3.3, one has to assume that the derivative P' has the α-property.

5. An application to partial differential equations

5.1. Consider the Dirichlet problem for the quasilinear equation

$$A(u) \equiv \sum_{|\alpha|=2m} a_\alpha(x,D^\gamma u)\cdot D^\alpha u + B(x,D^\beta u) = h(x) \;; \tag{5.1}$$

$$\gamma_j u \equiv D_n^j u|_\Gamma = \varphi_j(x') \;, \quad x' \in \Gamma \;, \quad (j = 0,1,\cdots,m-1) \;, \tag{5.2}$$

where $u(x) \in W^{(2m)}(\Omega)$, $h(x) \in L_p(\Omega)$, $\varphi_j(x') \in W_p^{(2m-j-1/p)}(\Gamma)$, $p > 1$, and $D_n^j u$ is the derivative of order j in the direction of the exterior normal $\bar{n}$ to the boundary Γ of the function $u(x)$; Ω is a bounded region of the space R^n with a sufficiently small boundary Γ. In order to apply the method of contractor directions, we reduce the Dirichlet problem to an operator equation in some Banach spaces, for instance, in $W_p^{(2m)}(\Omega)$, $W_p^{(2m-j-1/p)}(\Gamma)$, $j = 0,1,\cdots,m-1$, which are the Sobolev-Slobodeckii spaces of real functions; m is an integer (see Pohožaev [2]).

Suppose that the operator P defined by the formula

$$Pu \equiv \{A(u), \gamma_0 u, \cdots, \gamma_{m-1} u\} \tag{5.3}$$

acts from the real space $W_p^{(2m)}(\Omega)$ into the direct product

$$L_p(\Omega) \times \prod_{j=0}^{m-1} W_p^{(2m-j-1/p)}(\Gamma)$$

and is a continuous (or closed) Cauchy mapping and has a linear Gâteaux derivative operator $P'(u)$ for every $u \in W_p^{(2m)}(\Omega)$ which is defined by the formula

$$P'(u)v = \sum_{|\alpha|=2m} a_\alpha(x, D^\gamma u) D^\alpha v + \sum_{|\alpha|=2m} \sum_{\gamma\in M} a_{\alpha,\gamma}(x, D^\mu u) D^\gamma v \cdot D^\alpha u +$$

$$+ \sum_{|\beta|\leq 2m-1} B_\beta(x, D^\delta u) \cdot D^\beta v ,$$

where $a_{\alpha,\gamma} = \partial a_\alpha / \partial D^\gamma u$, $B_\beta = \partial B / \partial (B^\beta u)$, M is the range of the multi-index γ , $|\gamma| < 2m-n/p$ if $2mp > n$ and $M \neq 0$ for $2mp \leq n$, $\mu \in M$ and $|\delta| \leq 2m-1$.

Conditions given by Pohožaev [2] imply that hypotheses stronger than the above are satisfied.

Now we consider the linearized Dirichlet problem (5.4), (5.2), where

$$(5.4) \qquad P'(u)v = h , \quad h \in L_p(\Omega) ; \quad u,v \in W_p^{(2m)}(\Omega) .$$

We have to show that, for every $u(x) \in W_p^{(2m)}(\Omega)$ and any

$$(5.5) \qquad \{h(x), \varphi_0(x'), \cdots, \varphi_{n-1}(x')\} \in L_p(\Omega) \times \prod_{j=0}^{m-1} W_p^{(2m-j-1/p)}(\Gamma)$$

(of a dense subset), the linearized Dirichlet problem (5.4), (5.2) has a solution $v(x) \in W_p^{(2m)}$ which satisfies the condition (see condition (1.2))

$$\|v\|_{W_p^{(2m)}(\Omega)} \leq B(g(\|u\|_{W_p^{(2m)}(\Omega)}) \cdot (\|h\|_{L_p(\Omega)} + \sum_{j=0}^{m-1} \|\varphi_j\|_{W_p^{(2m-j-1/p)}(\Gamma)})),$$

where the function B satisfies conditions (i) - (iii) and g is a continuous function. Then the Dirichlet problem (5.1), (5.2) has a solution for any (5.5), by virtue of Theorem 1.4.

Now it is clear that the main difficulty lies in getting an a priori estimate for the linearized Dirichlet problem.

5.2. The following condition implies that the operator P defined by (5.3) is a Cauchy mapping:

a) If $\{u_n\}$ is any sequence of solutions

$$A(u_n) = h_n , \quad \gamma_j u_n = \varphi_j(x') , \quad n = 1,2,\cdots$$

of the Dirichlet problem (5.1), (5.2) with h replaced by h_n, then

$$\|u_n\|_{W_p^{(2m)}(\Omega)} \to \infty \quad \text{as} \quad n \to \infty$$

implies that

$$\|h_n\|_{L_p(\Omega)} + \sum_{j=0}^{m-1} \|\varphi_j^n\|_{W_p^{(2m-j-1/p)}(\Gamma)} \to \infty \quad \text{as} \quad n \to \infty .$$

The following condition b) given by Pohožaev [2] is sufficient for condition a) to be satisfied.

b) Suppose that there exists a real function $K(t)$ for nonnegative t, bounded for bounded values of the argument, such that for any possible solution $u(x) \in W_p^{(2m)}(\Omega)$ of the Dirichlet problem (5.1), (5.2) the inequality

$$\|u\|_{W_p^{(2m)}(\Omega)} \leq K(\|h\|_{L_p(\Omega)} + \sum_{j=0}^{m-1} \|\varphi_j\|_{W_p^{(2m-j-1/p)}(\Gamma)})$$

holds.

Let us notice that the argument used in our discussion is also valid for Hölder norms.

5.3. As an illustration to Theorem 1.6, consider the Dirichlet problem for the second order quasilinear equation with real coefficients:

$$\text{(5.6)} \qquad \begin{aligned} Pu \equiv - \sum_{k=1}^{N} \frac{\partial}{\partial x_k}(a_k(x,u,u_x)) - \sum_{k=1}^{N} c_k(x,u,u_x)\frac{\partial u}{\partial x_k} + \\ + c(x,u,u_x) + d(x,u,u_x) = f(x) , \end{aligned}$$

$$\text{(5.7)} \qquad u|_\Gamma = 0 , \quad \Gamma = \partial\Omega ,$$

where $x \in \Omega$ - a bounded domain in R^N; $\partial\Omega$ denotes the boundary of Ω; $f \in L^{2\alpha}(\Omega)$, $\alpha > 1$; $u \in H_0^{(1)}$ - the completion of $C_0^\infty(\Omega)$ in the norm $|u|_1$ defined by $|u|_1^2 = \int_\Omega \sum_{i=1}^{N} |\frac{\partial u}{\partial x_i}|^2 dx$. We assume that

$$\text{(5.8)} \qquad c_i(x,\bar{u},\bar{u}_1,\ldots,\bar{u}_N) = a_i(x,u,u_1,\ldots,u_N); \ \bar{u}_k = u_k \cdot p'/p ,$$

where $p(u) = (u^2+1)^{\beta}$, $\beta = (\alpha-1)/2\alpha$, $\alpha > 1$, $p' = dp/du$, $u_k = \partial u/\partial x_k$, $k = 1,2,\cdots,N$.

(5.9) $\qquad \partial a_k/\partial u = \partial c/\partial u_k$ and $\partial d/\partial u_k = c_k$ for $k = 1,\cdots,N$.

(5.10) $\qquad \partial d/\partial u = -\partial c/\partial u$ or $\partial d/\partial u \geq 0$ and $\partial c/\partial u \geq 0$.

It follows from (5.8) that

(5.11) $\quad p'\partial a_i/\partial u_k = p\partial c_i/\partial u_k$ and $p'\partial a_i/\partial u = p\partial c_i/\partial u$ for $i,k = 1,\cdots,N$.

We suppose that

(5.12)
$$\sum_{i,k=1}^{N} a_{ik}(x,u,u_1,\cdots,u_N)\xi_i\xi_k \geq C(u^2+1)^{-\beta}\sum_{i=1}^{N}\xi_i^2 ,$$

uniformly in $u_1,\cdots,u_N$, where $a_{ik} = \partial a_i/\partial u_k$, $i,k = 1,\cdots,N$; and C is a constant. The regularity conditions imposed on the coefficients are such that the operator P in (5.6) acts from the vector space $D(P) \subset H_0^{(1)}$ into $L^{2\alpha}(\Omega)$. Moreover, the operator P has a closed graph and a Gâteaux derivative defined as follows.

$$f = P'(u)h = -\sum_{i,k=1}^{N}\frac{\partial}{\partial x_k}\left(a_{ik}\frac{\partial h}{\partial x_i}\right) - \sum_{k=1}^{N}\frac{\partial}{\partial x_k}(b_k\cdot h) +$$

$$-\sum_{i,k=1}\frac{\partial u}{\partial x_k}\frac{\partial c_k}{\partial u_i}\frac{\partial h}{\partial x_i} - \sum_{k=1}^{N}\frac{\partial u}{\partial x_k}\frac{\partial c_i}{\partial u}h - \sum_{i=1}^{N}c_i\frac{\partial h}{\partial x_i} +$$

$$+\sum_{i=1}^{N}\frac{\partial c}{\partial u_i}\frac{\partial h}{\partial x_i} + \frac{\partial c}{\partial u}h + \sum_{i=1}^{N}\frac{\partial d}{\partial u_i}\frac{\partial h}{\partial x_i} + \frac{\partial d}{\partial u}h .$$

We have

$$-p\sum_{i,k=1}^{N}\left(a_{ik}\frac{\partial h}{\partial x_i}\right) = -\sum_{i,k=1}^{N}\left(pa_{ik}\frac{\partial h}{\partial x_i}\right) + p'\sum_{i,k=1}^{N}\frac{\partial u}{\partial x_k}a_{ik}\frac{\partial h}{\partial x_i} ;$$

$$-p\sum_{i=1}^{N}\frac{\partial}{\partial x_k}(b_k\cdot h) = -\sum_{k=1}^{N}\frac{\partial}{\partial x_k}(pb_kh) + p'\sum_{k=1}^{N}\frac{\partial u}{\partial x_k}b_k\cdot h ,$$

where $b_k = \partial c_k/\partial u$. Hence, we obtain, by virtue of (5.11) and (5.12),

$$(p\cdot f,h) = \int_\Omega pfh\,dx = (pP'(u)h,h) = \int_\Omega \sum_{i,k=1}^{N} pa_{ik}\frac{\partial h}{\partial x_i}\frac{\partial h}{\partial x_k}\,dx +$$

$$+ \int_\Omega p(\frac{\partial c}{\partial u} + \frac{\partial d}{\partial u})h^2\,dx \geq c\int_\Omega \sum_{i=1}^{N}(\frac{\partial h}{\partial x_i})^2 dx = c|h|_1^2 ,$$

by (9.10). Hence we obtain that

$$|h|_1 \leq c^{-1}(\int_\Omega p^2 f^2 dx)^{1/2} \leq c^{-1}(\int_\Omega (u^2+1)^{2\beta\gamma}\,dx)^{1/2\gamma}\cdot(\int_\Omega f^{2\alpha}dx)^{1/2\alpha}) ,$$

where $1/\alpha + 1/\gamma = 1$. Let $2\beta\gamma = 1$ or $\beta = (\alpha-1)/2\alpha$. Then we obtain

$$|h|_1 \leq c^{-1}(\int_\Omega (u^2+1)\,dx)^{1/2\gamma}\|f\|_{L^{2\alpha}(\Omega)} \leq$$

$$\leq c_0(\|u\|^2_{L^2(\Omega)}+1)^{1/2\gamma}\cdot\|f\|_{L^{2\alpha}(\Omega)} \leq c_0(\|u\|_{L^2}+1)\|f\|_{L^{2\alpha}} .$$

Hence, it follows that

$$|h|_1 \leq \bar{c}(|u|_1+1)\|f\|_{L^{2\alpha}(\Omega)} , \tag{5.13}$$

where $\bar{c}$ is some constant. Now suppose that, for arbitrary $u \in D(P)$, $P'(u)$ maps $D(P)$ onto a dense subset L_0 of $L^{2\alpha}(\Omega)$. Then the solution of the linearized equation $P'(u)h = f$, $f \in L_0$, satisfies condition (5.13).

Thus, it follows that the hypotheses of Theorem 1.6 are satisfied, and the Dirichlet problem (5.6), (5.7) has a solution for arbitrary $f \in L^{2\alpha}$, $\alpha > 1$.

5.2. Another example can be given without assumption (5.9) which is very restrictive. Let us consider the equation

$$Pu \equiv -\sum_{k=1}\frac{\partial}{\partial x_k}(a_k(x,u,u_x)) - \sum_{k=1}^{N} c_k(x,u,u_x) = f(x) , \tag{5.14}$$

where c_k $(k = 1,\cdots,N)$ are the same as in (5.6). Suppose that condition (5.12) is satisfied and in addition

$$|b_k(x,u,u_1,\dots,u_N) - c_k(x,u,u_1,\dots,u_N)| \le c_0/(u^2+1)^{\beta}\ , \tag{5.15}$$

where $b_k = \partial c_k/\partial u$ for $k = 1,\dots,N$; $\beta = (\alpha-1)/2\alpha$, $\alpha > 1$; and

$$C > c_0(\mathrm{diam}^2\Omega+1)/2\ , \tag{5.16}$$

where C is the constant in (5.12). The regularity conditions imposed on the coefficients are such that the operator P in (5.14) acts from the vector space $D(P) \subset H_0^{(1)}$ into $L^{2\alpha}(\Omega)$. Moreover, P has a closed graph and a Gâteaux derivative defined as follows.

$$f = P'(u)h = -\sum_{i,k=1}^{N} \frac{\partial}{\partial x_k}\left(a_{ik}\frac{\partial h}{\partial x_i}\right) - \sum_{k=1}^{N} \frac{\partial}{\partial x_k}(b_k\cdot h) + \tag{5.17}$$
$$- \sum_{i,k}^{N} \frac{\partial u}{\partial x_k}\frac{\partial c_k}{\partial u_i}\frac{\partial h}{\partial x_i} - \sum_{k=1}^{N}\frac{\partial u}{\partial x_k}\frac{\partial c_i}{\partial u}\cdot h - \sum_{i=1}^{N} c_i\frac{\partial h}{\partial x_i}\ .$$

By virtue of (5.11) and (5.12) we obtain for $p(u) = (u^2+1)^{\beta}$ that

$$(p\cdot f,h) = \int_\Omega pfh\,dx = (pP'(u)h,h) = \int_\Omega \sum_{i,k=1}^{N} pa_{ik}\frac{\partial h}{\partial x_i}\frac{\partial h}{\partial x_k}dx +$$

$$+ \int_\Omega ph\sum_{k=1}^{N}(b_k-c_k)\frac{\partial h}{\partial x_k}dx \ge C\sum_{k=1}^{N}\int_\Omega\left(\frac{\partial h}{\partial x_i}\right)^2 dx - 2^{-1}c_0\sum_{k=1}^{N}\int_\Omega\left[h^2+\left(\frac{\partial h}{\partial x_h}\right)^2\right]dx \ge$$

$$\ge C|h|_1^2 - 2^{-1}c_0(N\|h\|^2_{L^2(\Omega)} + |h|_1^2) \ge [C - c_0(\mathrm{diam}^2\Omega+1)/2]\,|h|_1^2 = c_1|h|_1^2$$

where c_1 is positive by virtue of (5.16). Hence, condition (5.13) follows in the same way as above. Suppose now that, for arbitrary $u \in D(P)$, $P'(u)$ maps $D(P)$ onto a dense subset L_0 of $L^{2\alpha}(\Omega)$. Then the solution of the linearized equation $P'(u)h = f$, $f \in L_0$, satisfies condition (5.13), and by virtue of Theorem 1.6, the Dirichlet problem (5.14), (5.7) has a solution for arbitrary $f \in L^{2\alpha}(\Omega)$.

5.3. Consider now the Dirichlet problem for the equation

$$-\sum_{k=1}^{N}\frac{\partial}{\partial x_k}(a_k(x,u,u_x)) + b(x,u,u_x) = 0 \tag{5.18}$$

with boundary condition (5.7). The following assumptions are needed.

(i) The operator $P: D(P) \subset H_0^{(1)} \to L^{2\alpha}(\Omega)$ is defined by (5.14) and satisfies conditions (5.12), (5.15) and (5.16). The graph of P is closed in $H_0^{(1)} \times L^{2\alpha}(\Omega)$. The operator P has a Gâteaux (additive and homogeneous) derivative which for fixed u is defined by (5.17) and has its range L_u dense in $L^{2\alpha}(\Omega)$.

(ii) If the Dirichlet problem (5.14), (5.7) has a solution, then it is unique.

(iii) The mapping $T: H_0^{(1)} \to L^{2\alpha}(\Omega)$, where $\alpha > 1$ and

$$Tu \equiv \sum_{k=1}^{N} c_k(x,u,u_x)\frac{\partial u}{\partial x_k} + b(x,u,u_x) \tag{5.19}$$

with c_k $(k = 1,\cdots,N)$ defined by (5.8), is a compact operator.

Finally,

(iv) There exists a constant K (a priori independent of t) such that if u is a solution of the equation

$$\sum_{k=1}^{N} \frac{\partial}{\partial x_k}(a_k(x,u,u_x)) - \sum_{i=1}^{N} c_i(x,u,u_x)\frac{\partial u}{\partial x_i} +$$

$$+ t\sum_{i=1}^{N} c_i(x,tu,tu_x) + b(x,tu,tu_x) = 0 , \quad 0 < t < 1 ,$$

with boundary condition (5.7), then $|u|_1 < K/t$.

Under these hypotheses the Dirichlet problem (5.18), (5.7) has a solution.

Proof. It follows from (i), by virtue of Theorem 1.6, that the operator P defined by (5.14) is a mapping onto $L^{2\alpha}(\Omega)$. Let $S = S(u_0,r) \subset H_0^{(1)}$ be an arbitrary open ball with center $u_0 \in D(P)$ and radius r, and consider the operator P on $D(P) \cap \overline{S}$, where $\overline{S}$ is the closure of S. It follows from Theorem 1.1a that if ρ_0 is sufficiently small, then for arbitrary $f \in S(Pu_0,\rho_0) \subset L^{2\alpha}(\Omega)$ there exists a solution $u \in D(P) \cap \overline{S}(u_0,r)$ of the equation $Pu = f$. By virtue of (ii), this solution is unique. Hence, it follows that the inverse mapping P^{-1} is continuous at Pu_0. Since P is onto, the continuity of the inverse mapping P^{-1} follows. Consider now the mapping $-P^{-1}T$ which is a compact operator since so is T, by assumption (iii). Let $\overline{S}(0,K) \subset H_0^{(1)}$ be the closed ball with center 0 and radius K defined in (iv). The Leray-Schauder degree theory can be applied to the mapping $-P^{-1}T:\overline{S}(0,K) \to H_0^{(1)}$. Then, by virtue of (iv), it follows that the operator $-P^{-1}T$ has a fixed point in $\overline{S}(0,K)$, i. e., there exists a function $u \in \overline{S}(0,K)$ such that $u = -P^{-1}Tu$ or $Pu + Tu = 0$. In other words, the Dirichlet problem (5.19), (5.7) has a generalized solution.

Solomyak [1] discussed the Dirichlet problem (5.19), (5.7) under different hypotheses by using the method of successive approximations. In particular, instead of (5.12) he considered the following algebraic inequality for the Dirichlet form

$$\sum_{i,k=1}^{N} a_{ik}(x,u,u_1,\cdots,u_N)\xi_i\xi_k \geq c(T^{n-2} + |u|^{n-2} + 1)\sum_{i=1}^{N}\xi_i^2 ,$$

where $T^2 = \sum_{i=1}^{N} u_i^2$; n is such that the function $f(x)$ which replaces 0 in (9.18) belongs to $L^p(\Omega)$ with $p = n/(n-1)$.

Remark 5.1. Suppose that in addition to the hypotheses (i) - (iv) the following assumption is made:

(v) There exists a real function $C(t)$ for nonnegative t, bounded on bounded sets of values of the variable t, such that, for any solution h, of the linear Dirichlet problem (5.17), (5.7), the inequality

$$\|h\|_{2\alpha} \leq C(|u|_1)\|f\|_{L^{2\alpha}(\Omega)}$$

holds, where $\|h\|_{2\alpha} = (\int_\Omega \sum_{i=1}^{N} |\frac{\partial h}{\partial x_i}|^{2\alpha} dx)^{1/2\alpha}$.

Then the Dirichlet problem (5.19), (5.7) has a solution u with $\|u\|_{2\alpha} < \infty$. In case (v), the norm $|\cdot|_1$ in the graph of P should be replaced by $\|\cdot\|_{2\alpha}$.

A similar statement also holds for the Sobolev norm $\|\cdot\|_{W_{2\alpha}^{(2)}(\Omega)}$,

$\alpha > 1$.

These assertions can be proved by introducing some changes in the proof of Theorem 1.6.

6. An estimate for the linearized Dirichlet problem

6.1. Koševlev [1] studied the Dirichlet problem for the second order linear differential equation

$$(6.1) \qquad L(u) \equiv \sum_{i,k=1} a_{ik}(x)\frac{\partial^2 u}{\partial x_i \partial x_k} + \sum_{j=1}^{N} b_j(x)\frac{\partial u}{\partial x_j} + c^2(x)u = f(x)$$

with boundary conditions

$$(6.2) \qquad u|_\Gamma = 0 ,$$

where $x = (x_1,\cdots,x_N) \in \Omega$, $\Gamma = \partial\Omega$ is the boundary of the bounded domain Ω in the N-dimensional Euclidean space with norm $|\cdot|$. Under the hypotheses that the coefficients $a_{ik}(x)$, $b_j(x)$ and $c(x)$ are continuous in $\Omega+\Gamma$ and $f \in L^p(\Omega)$, and for arbitrary numbers $\xi_1,\cdots,\xi_N$ $(p > 1)$ the following inequality is satisfied

$$\sum_{i,k=1}^{N} a_{ik}(x)\xi_i\xi_k \geq \lambda_0^2 \sum_{i=1}^{N} \xi_i^2 , \quad (\lambda_0 > 0) ,$$

he obtained the following important estimate for the solution u of the Dirichlet problem (6.1), (6.2),

$$\|u\|_{W_p^{(2)}(\Omega)} \leq c_1\|f\|_{L^p(\Omega)} + c_2 \sup_{\Omega_{N-1}} \int_{\Omega_{N-1}} |u|d\Omega_{N-1} ,$$

provided that the solution u exists, is bounded and $\|u\|_{W_p^{(2)}(\Omega)} < \infty$. The constants c_1 and c_2 depend on the moduli of continuity of the functions $a_{ik}(x)$ and the maxima of the moduli of the functions $b_j(x)$ and $c(x)$. The norm in the Sobolev space $W_p^{(2)}(\Omega)$ is defined as follows:

$$\|u\|_{W_p^{(2)}(\Omega)} = \sum_{i,k=1}^{N} \left\|\frac{\partial^2 u}{\partial x_i \partial x_k}\right\|_{L^p(\Omega)} + \|u\|_{L^p(\Omega)} .$$

Koševlev assumes that Γ is of the Liapunov type and hence it follows that near any point $P \in \Gamma$, the equation of the surface can be represented in the form:

$$y_N = \omega(y_1, y_2, \cdots, y_N) ,$$

where the direction of the y_n axis coincides with the outward normal direction to the surface and the axes $y_1, y_2, \cdots, y_{N-1}$ lie in the tangent manifold.

Koševelev's argument will be a basis for an estimate for the solution u of the linearized Dirichlet problem (6.3), (6.2)

$$\text{(6.3)} \quad \begin{aligned} L(z)u \equiv & \sum_{i,k=1}^{N} a_{ik}(x,z,z_x)\frac{\partial^2 u}{\partial x_i \partial x_k} + \sum_{j=1}^{N} b_j(x,z,z_x)\frac{\partial u}{\partial x_j} + c^2(x,z,z_x)u \\ = & f(x) , \end{aligned}$$

where $u, z \in W_p^{(2)}(\Omega)$, $z_x = (z_1, \cdots, z_N)$, $z_i = \partial z/\partial x_i$, $(i = 1, \cdots, N)$, $f \in L^p(\Omega)$. The real coefficients $a_{ik} = a_{ki}$ are continuously differentiable with respect to all variables. There exist functions β_j $(j = 1, \cdots, N)$ and γ which are continuous, nondecreasing in each variable and such that

$$\text{(6.4)} \quad \begin{aligned} & |b_j(x,z,z_1,\cdots,z_N)| \le \beta_j(|z|, |z_1|, \cdots, |z_N|) \\ & c^2(x,z,z_1,\cdots,z_N) \le \gamma(|z|, |z_1|, \cdots, |z_N|) , \end{aligned}$$

uniformly in $x \in \Omega + \Gamma$.

<u>Lemma 6.1 (Košelev [1])</u>. Consider the equation with constant coefficients

$$\sum_{i=1}^{N} \lambda_i^2 \frac{\partial^2 u}{\partial x_i^2} + \sum_{j=1}^{N} b_j \frac{\partial u}{\partial x_j} - c^2 u = f$$

inside the parallelepiped

$$\text{(6.5)} \quad Q_{r\pi\lambda} : 0 \le x_i \le r\lambda_i \pi ; \quad (0 < r < 1 ; \quad 0 < \lambda_0 < \lambda_i < 1) ,$$

where $u(x) = 0$ for $x \in \partial Q_{r\pi\lambda}$. The following estimate is true for the generalized solution u.

$$\|u\|_{W_p^{(2)}(Q_{r\pi\lambda})} \leq A\|f\|_{L^p(Q_{r\pi\lambda})},$$

where

$$A = C\lambda_0^{-2}KL^{-1} \max \exp\left(\frac{1}{2}\sum_{j=1}^{N} b_j x_j/\lambda_j^2\right),$$

$$K = 1 - \left[r\left(c^2 + \frac{1}{4}\sum_{j=1}^{N} b_j^2/\lambda_j^2\right)\right]^{2(k_0+1)}, \quad k_0 = [N/2], \quad 2k_0 + 1 > N,$$

$$L = 1 - \left[r\left(c^2 + \frac{1}{4}\sum_{j=1}^{N} b_j^2/\lambda_j^2\right)\right]^2, \quad C \text{ is an absolute constant.}$$

Lemma 6.2 (Košelev [1]). The generalized solution of the equation with constant coefficients

$$\sum_{i=1}^{N} \lambda_i \frac{\partial^2 u}{\partial x_i^2} + \sum_{j=1}^{N} b_j \frac{\partial u}{\partial x_j} - c^2 u = 0$$

and boundary conditions

$$u\Big|_{\substack{x_i = 0,\ \lambda_i\pi r \\ x_N = 0}} = 0 \quad (i = 1,2,\cdots,N-1) \ : \ u\Big|_{x_N = \lambda_N\pi r} = \varphi,$$

where φ is integrable on the edge $x_N = \lambda_N\pi r$ of the parallelepiped (6.5), exists for any domain $G \subset Q_{r\pi\lambda}$ with distance greater than d_0. The following inequality is valid.

$$\|u\|_{W_p^{(2)}(G)} \leq BD^{-1}\int_{x_N=\lambda_N\pi r} |\varphi| dx_1,\cdots,dx_{N-1},$$

where $B = C\lambda_0^{-1} \max \exp\left(\frac{1}{2}\sum_{j=1}^{N} b_j x_j/\lambda_j^2\right) FH^{-1}$, $F = r^2\left(c^2 + \frac{1}{4}\sum_{j=1}^{N} b_j^2/\lambda_j^2\right) + 1$;

$H = 1 - \exp\left(-2\left[4\left(c^2 + \frac{1}{4}\sum_{j=1}^{N} b_j^2/\lambda_j^2\right)^{1/2} + 1\right]\pi\right)$. $D = d_0^m r^m$, $0 < d_0 < 1$, $0 < r < 1$, where m is a constant depending only on N and p.

If condition (6.6) is replaced by

$$u|_{S_{r\pi\lambda}} = \varphi,$$

where φ is an integrable function defined on each of the edges $S_{r\pi\lambda}$, then for any interior domain G with distance from $S_{r\pi\lambda}$ greater than d_0, the following estimate holds

$$\|u\|_{W_p^{(2)}(G)} \leq CBD^{-1} \int_{Sr\pi\lambda} |\varphi| dS_{r\pi\lambda} ,$$

where C is an absolute constant, B and D are the same as above.

Lemma 6.3. Suppose that the coefficients $a_{ik}(x,z,z_1,\cdots,z_N)$ are continuously differentiable and there exist continuous nondecreasing functions p_{ik}, q_{ik}, r_{ik} such that the partial gradients of a_{ik} satisfy the following inequalities:

(i) $$|\text{grad}_x a_{ik}| \leq p_{ik}(|z|,|z_1|,\cdots,|z_N|) ;$$

(ii) $$|\text{grad}_z a_{ik}| \leq q_{ik}(|z|,|z_1|,\cdots,|z_N|) ;$$

(iii) $$|\text{grad}_{\bar{z}} a_{ik}| \leq r_{ik}(|z|,|z_1|,\cdots,|z_N|)$$

uniformly in $x \in \Omega$, where $\bar{z} = (z_1,z_2,\cdots,z_N)$.

Let $z = z(x)$, $x \in \Omega$ be an arbitrary but fixed function of $W_p^{(2)}(\Omega)$, $p > N$. Put $a_{ik}(x,z,z_x) = a_{ik}(x,z(x),\partial z/\partial x_1,\cdots,\partial z/\partial x_N)$, and $a_{ik}(x^0,z^0,z_x^0)$ indicates the value of a_{ik} at $x = x^0$.

Then there exists a continuous nondecreasing function q which has the following property: For arbitrary positive η, there is a positive r such that $|x-x_0| \leq r$ for all $x \in \omega$ implies

$$\sum_{i,k=1}^{N} \max_{x\in\omega} |a_{ik}(x,z,z_x) - a_{ik}(x^0,z^0,z_x^0)| \leq \eta \tag{6.7}$$

and

$$1/r^{1/\rho} \leq 2^{-1}\eta^{-1}\exp(q(\|z\|_{W_p^{(2)}(\Omega)})) , \tag{6.8}$$

where ω is an arbitrary domain such that $x^0 \in \omega \subset \Omega$, $\rho = (p-N)N/p(p+N)$.

Proof. We have $\text{Grad}\, a(x) = (da/dx_1, da/dx_2,\cdots,da/dx_N)$, where

$\frac{da}{dx_i} = \frac{\partial a}{\partial x_i} + \frac{\partial a}{\partial z}\frac{\partial z}{\partial x_i} + \sum_{j=1}^{N} \frac{\partial a}{\partial z_j}\frac{\partial^2 z}{\partial x_j \partial x_i}$. Hence, $a_{ik}(x,z,z_x) - a_{ik}(x^0,z^0,z_x^0) = \text{Grad}\, a_{ik}(x^0 + \theta(x-x^0))\cdot(x-x^0)$. By virtue of (i) - (iii) and Sobolev's [1] imbedding theorem, we obtain

$$\|a_{ik}(x,a,z_x) - a_{ik}(x^0,z^0,z_x^0)\|_{L^{\bar{p}}(\omega)} \leq C_1 r\cdot\exp(\bar{p}_{ik}(\|z\|_{W_p^{(2)}(\Omega)})) +$$

$$+ C_2 r\cdot\exp(\bar{q}_{ik}(\|z\|_{W_p^{(2)}(\Omega)}))\cdot\|z\|_{W_p^{(2)}(\Omega)} +$$

$$+ r\cdot\exp(\bar{r}_{ik}(\|z\|_{W_p^{(2)}(\Omega)}))\cdot\|z\|_{W_p^{(2)}(\Omega)} \leq$$

$$\leq r\cdot\exp(q^{ik}(\|z\|_{W_p^{(2)}(\Omega)}))$$

for some continuous nondecreasing function q^{ik} where C_1 and C_2 are absolute constants, and $\bar{p}_{ik}$, $\bar{q}_{ik}$, $\bar{r}_{ik}$ are continuous nondecreasing functions obtained from p_{ik} , q_{ik} , r_{ik} , repsectively, and $\bar{p} = (N+p)/2$.

Let us compute $I = (\sum_{j=1}^{N} \int_\omega |\frac{da_{ik}}{dx_j}|^{\bar{p}} dx)^{1/\bar{p}}$. We have

$$(\int_\omega |\frac{da_{ik}}{dx_\ell}|^{\bar{p}} dx)^{1/\bar{p}} \leq \bar{C}_1 r^{N/\bar{p}}.\exp(\bar{p}_{ik}(\|z\|_{W_p^{(2)}})) +$$

$$+ \bar{C}_2 r^{N/\bar{p}}.\exp(\bar{q}_{ik}(\|z\|_{W_p^{(2)}(\Omega)}))\cdot\|z\|_{W_p^{(2)}(\Omega)} +$$

$$+ \exp(\bar{r}_{ik}(\|z\|_{W_p^{(2)}(\Omega)})) \sum_{j=1}^{N} (\int_\omega |\frac{\partial^2 z}{\partial x_j \partial x_\ell}|^{\bar{p}} dx)^{1/\bar{p}} ,$$

and

$$(\int_\omega |\frac{\partial^2 z}{\partial x_j \partial x_\ell}|^{\bar{p}} dx)^{1/\bar{p}} \leq \|z\|_{W_{\alpha\bar{p}}^{(2)}(\Omega)}\cdot(2r)^{N/\beta\bar{p}} ,$$

where $\bar{p} = (N+p)/2$, $\alpha\bar{p} = p$, $\alpha^{-1} + \beta^{-1} = 1$, $\beta = 2p/(p-N)$, $N/\beta\bar{p} = (p-N)N/p(p+N) = 1/\rho$, $\bar{C}_1$, $\bar{C}_2$ are absolute constants, and $\bar{p}_{ik}$, $\bar{q}_{ik}$, $\bar{r}_{ik}$ are continuous nondecreasing functions obtained from p_{ik} , q_{ik} , r_{ik} , respectively. Hence,

$$I \leq r^{1/\rho} \exp(p^{ik}(\|z\|_{W_p^{(2)}(\Omega)})) ,$$

where p^{ik} is some continuous nondecreasing function.

Thus, by virtue of Sobolev's imbedding theorem, we obtain

$$\sum_{i,k=1}^{N} |a_{ik}(x,z,z_x) - a_{ik}(x^0,z^0,z_x^0)| \leq$$

$$\leq C \sum_{i,k=1}^{N} \|a_{ik}(x,z,z_x) - a_{ik}(x^0,z^0,z_x^0)\|_{W_{\bar{p}}^{(1)}(\omega)} \leq$$

$$\leq r^{1/\rho} \cdot \exp(q(\|z\|_{W_p^{(2)}(\Omega)})) ,$$

where q is a continuous nondecreasing function, C is an absolute constant. Suppose that the quadratic form for the linearized equation (6.3) satisfies the following conditions:

Let $x \in \Omega$, $z, z_1, \cdots, z_N$ be arbitrary. Then the eigenvalues satisfy the following inequalities

(iv) $$0 < \lambda_0^2 < \lambda_i^2 (x,z,z_1,\cdots,z_N) \leq 1$$

and there exists a continuous nondecreasing function $\bar{w}_0$ such that

(v) $$1/\lambda_0^2 \leq \bar{w}_0(|z|, |z_1|, \cdots, |z_N|) .$$

Let now z be an arbitrary element of $W_p^{(2)}(\Omega)$. Then there exists a continuous nondecreasing function w_0 such that

(6.9) $$1/\lambda_0^2 \leq w_0(\|z\|_{W_p^{(2)}(\Omega)})$$

and it follows from (6.4) that there is a continuous nondecreasing function w_1 such that

(6.10) $$c^2 + (4\lambda_0^2)^{-1} \sum_{j=1}^{N} b_j^2 \leq w_1(\|z\|_{W_p^{(2)}(\Omega)}) , \quad w_1(0) \geq 1 .$$

Under these hypotheses we can estimate the constant A in Lemma 6.1

assuming that the constant coefficients in question are obtained from equation (6.3) for fixed $x^0 \in \Omega$ and $z \in W_p^{(2)}(\Omega)$. We choose r so as to satisfy the inequality

$$r \cdot w_1(\|z\|_{W_p^{(2)}(\Omega)}) \leq 1/2 , \tag{6.11}$$

where w_1 is the same as in (6.10). Then we obtain

$$A \leq \exp(w_2(\|z\|_{W_p^{(2)}(\Omega)}) , \tag{6.12}$$

where $w_2 = w_0 + w_1 + \bar{c}$, $\bar{c}$ being an arbitrarily large absolute constant. Hence, it follows that K and L can be omitted and (6.11) does not contradict (6.8) under proper choice of η to be given below. The estimate (6.12) is also valid for B in Lemma 6.2, i.e.,

$$B \leq \exp(w_2(\|z\|_{W_p^{(2)}(\Omega)})) . \tag{6.13}$$

It remains to estimate $1/r$ in (6.8), where the proper choice for η which is required below is

$$0 < \eta = 1/2A2^{m+1}(2N/\lambda_0+1)^N , \tag{6.14}$$

with A , λ_0 replaced by estimates (6.12), (6.9). Then we obtain, by virtue of (6.8), (6.9), (6.12), (6.14),

$$\begin{aligned} 1/r &\leq 2^{(m+3)\rho}e^{N\rho}\exp(\rho(w_2+2N^2w_0+q)(\|z\|_{W_p^{(2)}(\Omega)})) \leq \\ &\leq \exp(w_3(\|z\|_{W_p^{(2)}(\Omega)})) , \end{aligned} \tag{6.15}$$

where w_3 is a continuous nondecreasing function. We can now obtain an estimate for the solution of the linearized Dirichlet problem (6.3), (6.2) which is based on the above estimates. This estimate is restricted to the interior points of Ω only.

<u>Theorem 6.1</u>. Suppose that the real coefficients $a_{ik} = a_{ki}$ of equation (6.3) are continuously differentiable and satisfy conditions

(i) - (iii) of Lemma 6.3. The coefficients b_j, c^2 are continuously differentiable and satisfy conditions similar to (i) - (iii) and (6.4). The eigenvalues corresponding to the quadratic form satisfy (iv), (v), and the Dirichlet problem (6.2), (6.3) has a solution.

$$u \in \overset{\circ}{W}{}_p^{(2)}(\Omega) \quad \text{and} \quad \|u\|_{W_p^{(2)}(\Omega)} < \infty . \tag{6.16}$$

Let $G \subset \Omega$ be an arbitrary domain with positive distance from the boundary $\partial\Omega$. Then there exists continuous nondecreasing functions σ_1 and σ_2 such that

$$\|u\|_{W_p^{(2)}(G)} \leq \exp(\sigma_1(\|z\|_{W_p^{(2)}(\Omega)}))\|f\|_{L^p(\Omega)} + \tag{6.17}$$

$$+ \exp(\sigma_2(\|z\|_{W_p^{(2)}(\Omega)}))\sup_{\Omega_{N-1}} \int_{\Omega_{N-1}} |u|d\Omega_{N-1} .$$

Proof. Let $x \in G$ be arbitrary and denote by $Q = Q(x)$ the cube with center x and sides of length $\alpha(x) = r_0 d(x)/N^{1/2}$, parallel to the coordinate axes, where $d(x)$ is the distance from x to the boundary $\partial\Omega$. Following Košelev [1] let us introduce the Schauder [1] function

$$q(x) = \|u\|_{W_p^{(2)}(Q)} d^m(x) .$$

Let $x^0 \in G$ be arbitrary but fixed, and denote by $\overline{Q}_0 = \overline{Q}(x^0)$ the cube with center x^0 and sides of length

$$\overline{\alpha}(x^0) = r_0 d(x^0)/\lambda_0 > \alpha(x^0) ,$$

where $0 < r_0 < 1$ and $r_0 < \lambda_0/2N$, where $\lambda_0 < \lambda_i(x^0, z(x^0), z_x^0)$ is defined by (iv). Then it follows from the definition of $q(x)$ that

$$\|u\|_{W_p^{(2)}(\overline{Q}_0)} \leq 2^{m+1}(2N/\lambda_0+1)^N \|u\|_{W_p^{(2)}(Q_0)} , \tag{6.18}$$

since the cube $\overline{Q}_0$ can be covered by $M < (2N/\lambda_0+1)^N$ cubes $Q(x^i)$ see Košelev [1]).

Let T_0 be the rectangular parallelepiped with sides of length

$$\beta_i = r_0 d(x^0)\lambda_i/\lambda_0 , \quad (i = 1,2,\cdots,N) , \tag{6.19}$$

and parallel to the principal axes of the quadratic form for (6.3) at $x = x^0$. Then we have (see Koševlev [1])

$$Q_0 \subset T_0 \subset \bar{Q}_0 . \tag{6.20}$$

Let us write the differential equation (6.3) in the form

$$L_0(u) = \sum_{i,k=1}^{N} a_{ik}(x^0,z^0,z_x^0)\frac{\partial^2 u}{\partial x_i \partial x_k} + \sum_{j=1}^{N} b_j(x^0,z^0,z_x^0)\frac{\partial u}{\partial x_j} +$$

$$+ c^2(x^0,z_x^0)u = f + F \tag{6.21}$$

where

$$F = \sum_{i,k=1}^{N} [a_{ik}(x^0,z^0,z_x^0) - a_{ik}(x,z,z_x)]\frac{\partial^2 u}{\partial x_i \partial x_k} + \sum_{j=1}^{N} [b_j(x^0,z^0,z_x^0)$$

$$- b_j(x,z,z_x)]\frac{\partial u}{\partial x_j} + [c^2(x^0,z^0,z_x^0) - c^2(x,z,z_x)]u$$

and x is in the interior of Q_0 . On the boundary S_0 of the parallelepiped T_0 , the solution u of (6.3) assumes its values

$$u|_{S_0} = \bar{u} .$$

A linear transformation of the independent variables in (6.3) by means of an orthogonal matrix yields

$$\bar{L}_0(u) = \sum_{i=1}^{N} \lambda_i^2(x^0,z^0,z_x^0)\frac{\partial^2 u}{\partial x_i^2} + \sum_{j=1}^{N} \bar{b}_j(x^0,z^0,z_x^0)\frac{\partial u}{\partial x_j} +$$

$$+ c^2(x^0,z^0,z_x^0)u = f + F ,$$

where the same notation is used for the new independent variables in the coordinate system determined by the principle axes of the quadratic form for (6.21). The solution of the equation (6.22) can be written as a sum of two functions: $u = u_1 + u_2$, where

$$\bar{L}_0(u_1) = f + F \quad \text{and} \quad u_1|_{S_0} = 0 , \quad S_0 = \partial T_0 ,$$

$$\bar{L}_0(u_2) = 0 \quad \text{and} \quad u_2|_{S_0} = \bar{u} .$$

It follows from Lemma 6.1 and (6.20) that

$$(6.23) \qquad \|u_1\|_{W_p^{(2)}(Q_0)} \leq \|u_1\|_{W_p^{(2)}(T_0)} \leq A\|f+F\|_{L^p(T_0)} .$$

It can be shown that by a proper choice of $r = r_0$ the estimate

$$(6.24) \qquad A\|F\|_{L^p(T_0)} \leq \frac{1}{2}\|u\|_{W_p^{(2)}(Q_0)}$$

is valid, where A is to be replaced by its estimate (6.12). In fact, by means of Sobolev's imbedding theorem, we obtain

$$\|F\|_{L^p(T_0)} \leq \sum_{i,k=1} \max_{x\in T_0} |a_{ik}(x^0,z,z_x) - a_{ik}(x,z,z_x)| \cdot \|u\|_{W_p^{(2)}(T_0)} +$$

$$+ C_1 \sum_{j=1}^{N} \max_{x\in T_0} |b_j(x^0,z^0,z_x^0) - b_j(x,z,z_x)| \cdot \|u\|_{W_p^{(2)}(T_0)} +$$

$$+ C_2 \max_{x\in T_0} |c^2(x^0,z^0,z_x^0) - b_j(x,z,z_x)| \cdot \|u\|_{W_p^{(2)}(T_0)} ,$$

where C_1 and C_2 are some constants. Now, estimates like (6.7) in Lemma 6.3 can be obtained for all terms in the right hand side of the last inequality. Thus, we can assume that the estimate (6.7) includes all terms involved in F. Since absolute constants can be included in the corresponding expressions, we can put $r = r_0$ for η determined by (6.14) with A replaced by its estimate (6.12), and we obtain the estimate (6.24), by virtue of (6.18), and also the estimate (6.15) follows. We observe that $\sum_{j=1}^{N} \bar{b}_j^2(x^0,z^0,z_x^0) = \sum_{j=1}^{N} b_j^2(x^0,z^0,z_x^0)$, due to the orthogonality of the transformation matrix. Thus, we have, by virtue of (6.4) and (6.18),

$$A\|f\|_{L^p(T_0)} \leq A\eta\|u\|_{W_p^{(2)}(T_0)} \leq \frac{1}{2}[2^{m+1}(2N/\lambda_0+1)^N]^{-1}\|u\|_{W_p^{(2)}(\bar{Q}_0)} \leq$$

$$\leq \frac{1}{2}\|u\|_{W_p^{(2)}(Q_0)}$$

Now we have, by (6.20), (6.23) and (6.24),

$$\|u\|_{W_p^{(2)}(Q_0)} \le \|u_1\|_{W_p^{(2)}(Q_0)} + \|u_2\|_{W_p^{(2)}(Q_0)} \le$$

$$\le A\|f\|_{L^p(T_0)} + \frac{1}{2}\|u\|_{W_p^{(2)}(Q_0)} + \|u_2\|_{W_p^{(2)}(T_0)} .$$

Hence, we obtain, by (6.12),

$$(6.25) \quad \|u\|_{W_p^{(2)}(Q_0)} \le 2 \exp(w_2(\|z\|_{W_p^{(2)}(\Omega)})\|f\|_{L^p(\bar{Q}_0)} + 2\|u_2\|_{W_p^{(2)}(T_0)} .$$

We now apply to Lemma 6.2 the estimates (6.13), (6.15) and use condition (6.16). Thus, we obtain

$$\|u_2\|_{W_p^{(2)}(T_0)} \le CBD^{-1}\int_{S_0} |u|dS_0 \le CBD^{-1}\sup_{\Omega_{N-1}}\int_{\Omega_{N-1}} |u|d\Omega_{N-1} \le$$

$$\le \bar{C}d_0^m \exp((w_2+\bar{m}w_3)(\|z\|_{W_p^{(2)}(\Omega)}))\sup_{\Omega_{N-1}}\int_{\Omega_{N-1}} |u|d\Omega_{N-1} ,$$

where $\bar{C}$, $\bar{m}$ are absolute constant.

Hence, we obtain by virtue of (6.25) that

$$\|u\|_{W_p^{(2)}(Q_0)} \le 2 \exp(w_2(\|z\|_{W_p^{(2)}(\Omega)})\|f\|_{L^p(\Omega)} +$$

$$+ 2\bar{C}d_0^m \exp((w_2+\bar{m}w_3)(\|z\|_{W_p^{(2)}(\Omega)}))\sup_{\Omega_{N-1}}\int_{\Omega_{N-1}} |u|d\Omega_{N-1} .$$

The domain G can be covered by a finite number t of small cubes Q_i $(i = 1,\cdots,t)$. It is clear that t does not exceed C_0/r^N , where C_0 is an absolute constant. Hence, using the last estimate for $\|u\|_{W_p^{(2)}(Q_0)}$ and the estimate (6.15) for $1/r$, we obtain (6.17).

Remark 6.1. The exponential polynomial form of the factor in the estimate (6.17) is characteristic for elliptic equations even in the linear case (see Lemmas 6.1 and 6.2). Therefore, Theorem 1.6 which has been applied to some particular cases in Section 5, is not satisfactory in general case. In order to investigate the general case,

Theorem 1.4 can be used with $B(s) = s$, i.e., the following condition should be satisfied.

$$\|h\| \leq w(\|x\|)\,\|y\| \,, \tag{6.26}$$

where w is an arbitrary continuous function. Condition (α) is then required.

6.2. Consider the Dirichlet problem for the equation

$$Pz \equiv -\sum_{k=1}^{N} \frac{\partial}{\partial x_k}(a_k(x,z,z_x)) - c(x,z,z_x) = f \tag{6.27}$$

with boundary conditions (6.2). The following hypotheses are required.

(i) The differential operator

$$P : D(P) \subset \overset{\circ}{W}{}_p^{(2)}(\Omega) \to L^p(\Omega) \,, \quad p > N \,,$$

where $\overset{\circ}{W}{}_p^{(2)}(\Omega)$ is the Sobolev space of functions with compact supports, has a closed graph and satisfies condition (α). There exists the Gâteaux (additive and homogeneous) derivative P' which for fixed $z \in \overset{\circ}{W}{}_p^{(2)}(\Omega)$ is defined in (6.28)

$$\begin{aligned} -P'(z)u = {} & \sum_{i,k=1} \frac{\partial}{\partial x_k}\left(a_{ik}\frac{\partial u}{\partial x_i}\right) - \sum_{k=1}^{N} \frac{\partial}{\partial x_k}(b_k \cdot u) + \\ & + - \sum_{j=1}^{N} c_j \frac{\partial u}{\partial x_j} - bu = f \,, \end{aligned} \tag{6.28}$$

where $c_j = \partial c/\partial z_j$, $z_j = \partial z/\partial x_j$, $b = \partial c/\partial z$, $b_k = \partial a_k/\partial z$, $a_{ik} = a_{ki} = \partial a_k/\partial z_i$, $(i,j,k = 1,\cdots,N)$. Suppose that the quadratic form satisfies the inequality

$$\sum_{i,k=1}^{N} a_{ik}(x,z,z_1,\cdots,z_N)\xi_i\xi_k \geq C\bar{w}_0(|z|,|z_1|,\cdots,|z_N|)^{-1}\sum_{i=1}^{N}\xi_i^2 \,, \tag{6.29}$$

uniformly in $x \in \Omega$, where C is a constant and $\bar{w}_0$ is a continuous function nondecreasing in each variable and $\bar{w}_0(0) > 0$. We assume that

$$|b_k(x,z,z_1,\dots,z_N) - c_k(x,z,z_1,\dots,z_N)| \le \tag{6.30}$$
$$\le c_0/\bar{w}_0(|z|,|z_1|,\dots,|z_N|) ,$$

uniformly in $x \in \Omega$, and

$$c > c_0(\mathrm{diam}^2\Omega + 1)/2 . \tag{6.31}$$

The coefficients a_{ik} , c are continuously differentiable and the regularity hypotheses of Theorem 6.1 are satisfied for $P'(z)u$ in (6.28), and $b \le 0$. Then the Dirichlet problem (6.27), (6.2) has a generalized solution $z \in \overset{\circ}{W}{}^{(2)}_p(G)$ for any domain $G \subset \Omega$ with positive distance from the boundary Γ of Ω .

<u>Proof</u>. By virtue of (6.29) - (6.31) and Sobolev's imbedding theorem, we have, for arbitrary domain $\omega \subset \Omega$; $z,u \in \overset{\circ}{W}{}^{(2)}_p$

$$\int_\Omega fu\,dx = \int_\Omega u\cdot P'(z)u\,dx = \int_\Omega \sum_{i,k=1}^N a_{ik}\frac{\partial u}{\partial x_i}\frac{\partial u}{\partial x_k} +$$

$$+ \int_\Omega \sum_{k=1}^N (b_k - c_k)\frac{\partial u}{\partial x_k} u\,dx - \int_\Omega cu^2\,dx \ge$$

$$\ge w_0(\|z\|_{\overset{\circ}{W}{}^{(2)}_p(\Omega)})^{-1}\{c\int_\Omega \sum_{k=1}^N (\frac{\partial u}{\partial x_k})^2\,dx - 2^{-1}c_0 \sum_{k=1}^N \int_\Omega [u^2 + (\frac{\partial u}{\partial x_k})^2]dx\} \ge$$

$$\ge w_0(\|z\|_{W^{(2)}_p(\Omega)})^{-1}[c - c_0(\mathrm{diam}^2\Omega + 1)/2] \sum_{k=1}^N \int_\Omega (\partial u/\partial x_k)^2\,dx .$$

Hence,

$$\int_\Omega fu\,dx \ge c_1 w_0(\|z\|_{W^{(2)}_p(\Omega)})^{-1} \sum_{k=1}^N \int_\Sigma (\partial u/\partial x_k)^2 dx , \tag{6.32}$$

where $c_1 > 0$, by (6.31), and w_0 is the same as in (6.9). It follows from (6.32) and Sobolev's imbedding theorem that

$$\|u\|^2_{L^{p'}(\Omega_{N-1})} \le c\int_\Omega |f|\cdot|u|dx .$$

But

$$\int_\Omega |f|\cdot|u|dx = \int_{x_N} dx_N \int_{\Omega_{N-1}} |f|\cdot|u|d\Omega_{N-1} \le$$

$$\leq \int_{x_n} dx_n \left(\int_{\Omega_{N-1}} |f|^p d\Omega_{N-1}\right)^{1/p} \cdot \int_{\Omega_{N-1}} |u|^{p'} d\Omega_{N-1}\Big)^{1/p'}$$

$$\leq c\|u\|_{L^{p'}(\Omega_{N-1})} \|f\|_{L^p(\Omega)} .$$

Finally, we obtain

$$\sup_{\Omega_{N-1}} \int_{\Omega_{N-1}} |u| d\Omega_{N-1} \leq c\|f\|_{L^p(\Omega)} . \tag{6.33}$$

Combining (6.33) and (6.17), we obtain for some continuous nondecreasing function σ that

$$\|u\|_{W_p^{(2)}(G)} \leq \exp(\sigma(\|z\|_{W_p^{(2)}(\Omega)}))\|f\|_{L^p(\Omega)} . \tag{6.34}$$

It follows from Košelev's [1] theorem that the linearized Dirichlet problem (6.28), (6.2) has a solution $u \in \overset{\circ}{W}{}_p^{(2)}(\Omega)$ for $z \in \overset{\circ}{W}{}_p^{(2)}(\Omega)$ and for $f \in L^p(\Omega)$ under the assumption made above that $p > N$. Thus, the hypotheses of Theorem 6.1 are satisfied and the proof follows from Remark 6.1, by virtue of (6.34).

7. An application of contractor directions to nonlinear integral equations

As another illustration of the method of contractor directions, let us consider the following nonlinear integral equation

$$x(s) - \int_0^1 K(s,t,x(t))dt = f(s) , \tag{7.1}$$

where K is a continuous function. We introduce the nonlinear operator

$$y = Px , \quad y(s) = x(s) - \int_0^1 K(s,t,x(t))dt \tag{7.2}$$

and consider the integral equation (7.1) as an operator equation in $X = Y = C$

$$Px = f , \quad x,f \in C = C(0,1) , \tag{7.3}$$

where C is the Banach space of continuous functions with norm $\|x\| = \max_{0\leq t\leq 1} |x(t)|$. We assume, in general case, that P is a Cauchy mapping of C . Suppose now that the partial derivative $K'_x(s,t,x)$ exists and is continuous and let us consider the linearized integral equation

$$(7.4) \qquad h(s) - \int_0^1 K'_x(s,t,x(t))h(t)dt = f(s)$$

or in operator form

$$(7.5) \qquad P'(x)h = f \ , \quad x,h,f \in C \ ,$$

where $P'(x)h$, for fixed $x \in C$, is defined by the formula

$$(7.6) \qquad P'(x)h(s) = h(s) - \int_0^1 K_x(s,t,x(t))h(t)dt \ .$$

For fixed $x \in C$, put

$$(7.7) \qquad K_x(s,t) = K'_x(s,t,x(t))$$

and suppose that the kernel in (7.7) has a resolvent kernel $G_x(s,t)$ with the following property. There exists a continuous function g such that

$$(7.8) \qquad \int_0^1 |G_x(s,t)|dt \leq g(\|x\|) \quad \text{for all} \quad x \in C \ , \quad 0 \leq s \leq 1 \ .$$

Finally, assume that for each $x \in C$, the linearized equation (7.4) has a solution h for arbitrary $f \in C_x \subset C$, where C_x is some dense subset of C . Under these hypotheses, the integral equation (7.1) has a solution for arbitrary $f \in C$.

Proof. The proof follows from Theorem 1.4. In fact, we have, by virtue of (7.7), for the solution h of equation (7.4)

$$(7.9) \qquad h(s) = f(s) - \int_0^1 G_x(s,t)f(t)dt \ .$$

It results from (7.9) and (7.8) that

$$(7.10) \qquad \|h\| \leq (1 + g(\|x\|))\|f\| \ .$$

Hence, it follows that the hypotheses of Theorem 1.4 are satisfied with $B(s) = s$.

Example. Consider the nonlinear integral equation

$$(7.11) \qquad x(s) - \int_0^1 st[x(t) - \arctan x(t)]dt = f(s)$$

as a nonlinear operator equation (7.3), where the operator values $y = Px$ are determined by the formula

$$y(s) = x(s) + \int_0^1 st[x(t) - \arctan x(t)]dt .$$

The linearized integral equation is of the form

$$(7.12) \qquad P'(x)h(s) = h(s) + \int_0^1 st[1 - 1/[1+x^2(t)]]h(t)dt = f(s) .$$

Since the integral operator in (7.12) is a contraction for every fixed $0 \neq x \in C$, it follows that the linearized integral equation (7.12) has a (unique) solution for arbitrary $f \in C$. Furthermore, if $h \in C$ is a solution of equation (7.12), then we obtain that

$$|h(s)| - [1 - 1/(1+\|x\|^2)]\|h\| \leq |f(s)| \quad \text{for} \quad 0 \leq s \leq 1 .$$

Hence, it follows that

$$\|h\| - [1 - 1/(1 + \|x\|^2)]\|h\| \leq \|f\|$$

or

$$(7.13) \qquad (1 + \|x\|^2)^{-1}\|h\| \leq \|f\| .$$

Thus, the operator P determined by (7.11) is (B,g)-differentiable with $B(s) = s$ and $g(t) = 1 + t^2$, by virtue of (7.13).

CHAPTER 7

CONTRACTORS AND SECANT METHODS FOR SOLVING EQUATIONS

Introduction. In this chapter, a further development of the contractor concept is presented. Once the contractor technique is available, an analysis of the two point secant methods makes it clear how to generalize the concept of a contractor for the sake of handling these methods. In this way, the notion of a two point contractor is introduced. A two point contractor is actually a contractor associated with two variable points of a Banach space whereas a (simple) contractor is associated with one variable point only. The abstract two point secant method is based on the analog of the first divided difference for the operator involved in the equation. The divided difference operator is supposed to be nonsingular. This is usually an assumption made in case of two point secant methods. It follows from the definition that the inverse of a divided difference operator is a two point contractor. Therefore, the notion of a two point contractor can be considered as a generalization of an inverse of a divided difference operator. We describe a broad class of iterative methods based upon the notion of two point contractors. Various two point secant methods are obtained here as special cases. By using the two point contractor method existence theorems for solutions of equations are obtained as well as convergence theorems for the iterative methods in question. Moreover, the character of the convergence itself is also investigated. Under suitable assumptions the convergence turns out to be quadratic. It is known for various secant methods of Steffensen type that their convergence is quadratic. Thus, the two point secant methods are generalized without loss of the convergence rate.

2. Divided difference operators and two point contractors

Given two Banach spaces X and Y, let us denote by $L(Y \to X)$ the set of all linear bounded operators from Y to X. Let $P:X \to Y$ be a nonlinear operator and let $Q(s,t) \geq 0$ for $s,t \geq 0$ be a continuous real valued function.

Definition 2.1. A mapping $\Gamma:X \times X \to L(Y \to X)$ is said to be a two point contractor for P with majorant function Q if the following inequality is satisfied

$$\|P(x+\Gamma(x,\bar{x})y) - Px - y\| \leq Q(\|y\|, \|x-\bar{x}\|) \tag{2.1}$$

for $x,\bar{x} \in X$, $y \in Y$, whenever $x+\Gamma(x,\bar{x})y \in D$ - the domain of P. In applications the variables x and $\bar{x}$ usually run over balls in X while y lies in a ball of Y with center 0. Inequality (2.1) is called the contractor inequality.

Let us recall that a contractor as defined in Chapter 1, is a mapping $\Gamma:X \to L(Y \to X)$. Thus, the concept of a two point contractor generalizes the notion of a contractor. The following contractor inequality can be considered as a special case of (2.1).

$$\|P(x+\Gamma(x,\bar{x})y) - Px - y\| \leq o(\|y\|) + K\|x-\bar{x}\| , \tag{2.1*}$$

where K is a constant and $0 \leq o(s)/s \to 0$ as $s \to 0$, $s \geq 0$. It follows from (2.1*) that $\tilde{\Gamma}(x) = \Gamma(x,x)$ is an inverse Fréchet derivative as defined in Chapter 1, and, consequently, a contractor.

Following Schmidt [1], the first divided difference operator of P is a mapping

$$F:D \times D \subset X \times X \to L(X \to Y)$$

which satisfies the relationship $F(x,\bar{x})(x-\bar{x}) = Px - P\bar{x}$ for $x,\bar{x} \in D \subset X$. Suppose that $F(x,\bar{x})$ is nonsingular and put $\Gamma(x,\bar{x}) = F(x,\bar{x})^{-1}$, then the following is true:

Lemma 2.1. If $\|F(x,\bar{x})^{-1}\| \leq B$ for some constant B and

(2.2) $$\|F(x,\bar{x}) - F(\xi,\bar{\xi})\| \le K[\|x-\xi\| + \|\bar{x}-\bar{\xi}\|]$$

for $x,\bar{x},\xi,\bar{\xi} \in D$, then $\Gamma(x,\bar{x})$ is a two point contractor.

Proof. Put $h = \Gamma(x,\bar{x})y$, $y \in Y$; then we obtain

$$\|P(x+h) - Px - y\| = \|P(x+h) - Px - F(x,\bar{x})h\| =$$

$$= \|F(x+h)h - F(x,\bar{x})h\| \le K[\|h\| + \|x-\bar{x}\|]\|h\| \le$$

$$\le KB(B\|y\| + \|x-\bar{x}\|)\|y\| = Q(\|y\|, \|x-\bar{x}\|) .$$

Hence, under the hypotheses given above, the inverse of the first divided difference operator is a two point contractor. Note that the same is true if the first divided difference operator satisfies the following condition instead of (2.2)

(2.2*) $$\|Px - P\bar{x} - F(\xi,\bar{x})(x-\bar{x})\| \le K\|x-\bar{x}\| \cdot \|x-\xi\| .$$

In fact, we have for $h = \Gamma(x,\bar{x})y$,

$$\|P(x+h) - Px - y\| = \|P(x+h) - Px - F(x,\bar{x})h\| \le$$

$$\le K\|h\| \cdot \|x+h-\bar{x}\| \le K\|h\|(\|h\| + \|x-\bar{x}\|) \le$$

$$\le KB\|y\|(B\|y\| + \|x-\bar{x}\|) = Q(\|y\|, \|x-\bar{x}\|) .$$

In this case again $\Gamma(x) = \Gamma(x,x) = F(x,x)^{-1}$ is an inverse Fréchet derivative and, consequently, a contractor.

A two point contractor is said to be bounded if there exists a positive constant B such that

(2.3) $$\|\Gamma(x,\bar{x})\| \le B \quad \text{for} \quad x,\bar{x} \in D \subset X .$$

Consider the operator equation

(2.4) $$Px = 0 ,$$

where $P: D \subset X \to Y$ is a nonlinear closed operator, i.e.,

$$x_n \in D , \quad x_n \to x \quad \text{and} \quad Px_n \to y$$

imply $x \in D$ and $y = Px$. We assume that P has a two point contractor Γ satisfying the contractor inequality (2.1) with $Q(s,t)$ continuous and nondecreasing in each variable and $Q(0,0) = 0$. In order to solve (2.4), we consider the following iterative procedure

$$(2.5) \qquad x_{n+1} = x_n - \Gamma(x_n, \bar{x}_n) Px_n \ , \quad n = 0,1,\cdots \ ,$$

where $\bar{x}_n$ will be defined below. We consider also the iterative sequence $\{t_n\}$ defined as follows

$$(2.6) \qquad t_{n+1} = t_n + Q(t_n - t_{n-1}, C(t_n - t_{n-1})) \ , \quad n = 0,1,\cdots$$

with initial values $t_0 = 0$, $t_1 = Q(\eta, C\eta)$, where the constant C will be defined below.

<u>Theorem 2.1</u>. Let $P: D \subset X \rightarrow Y$ be a closed nonlinear operator with domain D containing the ball $S(x_0, r)$, where x_0 is such that $\|Px_0\| \le \eta$. Suppose that P has a two point contractor satisfying (2.3) and the contractor inequality (2.1) for $x, \bar{x} \in S$ whenever $x + \Gamma(x,\bar{x})y \in S$. Let $\bar{x}_n$ in (2.5) be chosen so as to satisfy

$$(2.7) \qquad \|x_n - \bar{x}_n\| \le C\|Px_n\| \ , \quad n = 0,1,\cdots \ .$$

Assume also that the sequence $\{t_n\}$ defined by (2.6) has a limit t^* and $r = Bt^*$. Then the sequence $\{x_n\}$ defined by (2.5) and (2.7) converges toward a solution x of (2.4); all x_n lie in S and the error estimate holds

$$(2.8) \qquad \|x - x_n\| \le B(t^* - t_n) \ .$$

<u>Proof</u>. It follows from the contractor inequality (2.1) with $y = -Px_n$, $x = x_n$, $\bar{x} = \bar{x}_n$, and by virtue of (2.5) using also (2.7) that

$$(2.9) \qquad \|Px_{n+1}\| \le Q(\|Px_n\|, C\|Px_n\|) \ , \quad n = 0,1,\cdots \ .$$

Now we show by induction that

$$(2.10_n) \qquad \|Px_n\| \le t_n - t_{n-1} \quad \text{for} \quad n = 1,2,\cdots \ .$$

In fact, using (2.9) with $n = 0$ we obtain

$$\|Px_1\| \leq Q(\|Px_0\|, C\|Px_0\|) \leq Q(\eta, C\eta) = t_1 - t_0$$

by virtue of (2.6). Assuming (2.10_n) and using (2.9) again we conclude by virtue of (2.6) that

$$\|Px_{n+1}\| \leq Q(t_n - t_{n-1}, C(t_n - t_{n-1})) = t_{n+1} - t_n$$

whence (2.10_{n+1}) follows.

Furthermore, (2.3), (2.5) and (2.10_n) imply

$$\|x_{n+m} - x_n\| \leq \sum_{i=n}^{n+m-1} \|x_{i+1} - x_i\| \leq B \sum_{i=n}^{n+m-1} \|Px_i\| \leq B \sum_{i=n}^{n+m-1} (t_{i+1} - t_i) \leq B(t_{n+m} - t_n) ,$$

that is

$$\|x_{n+m} - x_n\| \leq B(t_{n+m} - t_n) . \tag{2.11}$$

Inequality (2.11) shows that $\{x_n\}$ is a Cauchy sequence. Denote by x its limit. Since $Px_n \to C$ as $n \to \infty$ as by (2.10_n) and P is closed we infer that $Px = 0$. Putting $n = 0$ in (2.11), we obtain

$$\|x_m - x_0\| \leq Bt_m < Bt^* \quad \text{for} \quad m = 1, 2, \cdots .$$

Letting $m \to \infty$ in (2.11) we obtain the error estimate (2.8) and the proof is complete.

Let us discuss some special cases of condition (2.7) in conjunction with procedure (2.5).

Consider the procedure (2.5) where $\bar{x}_n$ is defined as follows

$$\bar{x}_{n+1} = x_{n+1} - \Gamma(x_n, \bar{x}_n) Px_{n+1} , \quad n = 0, 1, \cdots . \tag{2.12}$$

Assuming (2.3) we infer from (2.12) that condition (2.7) is satisfied with $C = B$. Suppose now that the majorant function Q has the form $Q(s,t) = KB(Bs + t)s$, where B , K are some positive constants. Thus the contractor inequality (2.1) becomes

$$\|P(x + \Gamma(x, \bar{x})y) - Px - y\| \leq BK(B\|y\| + \|x - \bar{x}\|)\|y\| . \tag{2.13}$$

In this case the procedure defined by (2.5), (2.12) is quadratically convergent and we have the following

Theorem 2.2. Let $P:D \subset X \to Y$ be a closed nonlinear operator with domain D containing the ball $S = S(x_0,r)$, where $\|Px_0\| \leq \eta$, $\|x_0-\bar{x}_0\| \leq B\eta$, $a = 2KB^2$, $q = a\eta < 1$, $r = B\eta t^*$ and $t^* = \sum_{n=0}^{\infty} q^{2^n-1}$

Suppose that P has a two point bounded contractor Γ satisfying (2.3) and the contractor inequality (2.13) for $x,\bar{x} \in S$ whenever $x + \Gamma(x,\bar{x})y \in S$.

Then the sequence $\{x_n\}$ defined by (2.5), (2.12) converges to a solution x of the equation $Px = 0$. All x_n lie in S and the error estimate holds:

$$\|x-x_n\| \leq B\eta t^* q^{2^n-1} . \tag{2.14}$$

Proof. It follows from the contractor inequality (2.13) with $x = x_n$, $\bar{x} = \bar{x}_n$ and $y = -Px_n$ that

$$\|Px_{n+1}\| \leq K(1+B)\|Px_n\|^2 \leq at_n^2 = t_{n+1} , \tag{2.15}$$

where $t_0 = \eta$, $n = 1,2,\cdots$ and $\|Px_1\| \leq a\eta^2 = t_1$. We prove by induction that

$$t_n = \eta q^{2^n-1} \quad \text{for} \quad t_{n+1} = at_n^2 , \quad n = 0,1,\cdots . \tag{2.16}$$

Hence, by virtue of (2.5), (2.3), (2.15) and (2.16) we obtain

$$\|x_{n+1}-x_n\| \leq B\eta q^{2^n-1} ,$$

$$\|x_{n+1}-x_0\| \leq B\eta \sum_{i=0}^{n} q^{2^i-1} < B\eta t^* .$$

The latter inequality shows that all x_n lie in S. Further, we have

$$\|x_{n+m}-x_n\| \leq B\eta \sum_{i=n}^{n+m-1} q^{2^i-1} < B\eta t^* q^{2^n-1} . \tag{2.17}$$

Thus, $\{x_n\}$ is a Cauchy sequence and has a limit x. It follows from (2.15), (2.16) and $q < 1$ that $Px_n \to 0$ as $n \to \infty$. Since P is closed, we conclude that $Px = 0$. The error estimate (2.14) follows from (2.17) by letting $m \to \infty$. We assume now that the first divided difference operator F for P exists and is nonsingular. Then we consider the following iterative procedure investigated by Laasonen [1].

$$x_{n+1} = x_n - F(x_n,\bar{x}_n)^{-1}Px_n \tag{2.18}$$

$$\bar{x}_{n+1} = x_{n+1} - F(x_n,\bar{x}_n)^{-1}Px_n \,, \tag{2.19}$$

for $n = 0,1,\cdots$.

Suppose, in addition, that condition (2.2) or (2.2*) is satisfied on $S = S(x_0,r) \subset D$.

Under these hypotheses Lemma 2.1 shows that $\Gamma(x,\bar{x}) = F(x,\bar{x})^{-1}$ is a two point bounded contractor for P. Thus, we can apply Theorem 2.2 to the procedure (2.18), (2.19) and we obtain the following

<u>Theorem 2.3</u>. Let $P:D \subset X \to Y$ be a closed nonlinear operator with domain D containing the ball $S = S(x_0,r)$. Suppose that there exists the first divided difference F for P satisfying (2.2) or (2.2*) on S, F is nonsingular and

$$\|F(x,\bar{x})^{-1}\| \leq B \quad \text{for} \quad x,\bar{x} \in S \,. \tag{2.20}$$

The initial elements x_0 and $\bar{x}_0$ are chosen so as to satisfy

$\|Px_0\| \leq \eta$, $\|x_0-\bar{x}_0\| \leq B\eta$, $r = B\eta t^*$, $a = 2KB^2$, $q = a\eta < 1$ and $t^* = \sum_{n=0}^{\infty} q^{2^n-1}$.

Then the sequence $\{x_n\}$ defined by (2.18), (2.19) converges to a solution x of the equation $Px = 0$. All x_n lie in S and the error estimate (2.14) holds.

<u>Proof</u>. It follows from the hypotheses of the theorem by virtue of Lemma 2.1 that $\Gamma(x,\bar{x}) = F(x,\bar{x})^{-1}$ is a bounded two point contractor satisfying the contractor inequality (2.13). It is easy to verify that all hypotheses of Theorem 2.2 are satisfied. Hence, the proof follows

immediately.

Remark 2.1. The iterative procedure (2.5), (2.13) generalizes the procedure (2.18), (2.19) considered by Laasonen [1] who uses a different argument in his convergence proof and error estimate. But our argument seems to be more general and the arithmetic used here is rather simpler.

Let us note that condition (2.20) for $x = x_n$ and $\bar{x} = \bar{x}_n$ which is important in our considerations is shown there to be also satisfied (see Laasonen [1]).

Let us observe that Theorem 2.3 generalizes Theorem 3.4, Chapter 2. In fact, put $\Gamma(x) = \Gamma(x,x)$. Then the contractor inequality (2.13) yields the corresponding contractor inequality for the contractor $\Gamma(x)$ as required in Theorem 3.4, Chapter 2. Obviously, we obtain the contractor method considered there by simply putting $\bar{x}_n = x_n$ in (2.12). As to Theorem 3.4, Chapter 2, it is shown there that it generalizes a theorem of Mysovskih [1] concerning the general Newton method.

3. Some particular cases of two point secant methods

Let us discuss other special cases of condition (2.7) in conjunction with procedure (2.5). Consider the procedure (2.5) where $\bar{x}_n$ is defined as follows

$$\bar{x}_n = x_n + c_n P x_n \ ; \ |c_n| \leq C \ ; \ n = 0,1,\cdots , \tag{3.1}$$

where C is some positive constant. It is clear from (3.1) that condition (2.7) is satisfied. As in Section 2, we assume that the majorant function Q has the form $Q(s,t) = BK(Bs+t)s$, where K , B are some positive constants. Thus, the contractor inequality (2.1) has the form (2.13). In this case the procedure defined by (2.5) and (3.1) is also quadratically convergent and we obtain the following

Theorem 3.1. Let $P:D \subset X \to Y$ be a closed nonlinear operator with domain D containing the ball $S = S(x_0,r)$, where $\|Px_0\| \leq \eta$, $a = BK(B+C)$, $q = a\eta < 1$, $r = B\eta t^*$ and $t^* = \sum_{n=0}^{\infty} q^{2^n-1}$. Suppose that P has a two point bounded contractor Γ satisfying (2.3) and the contractor inequality (2.13) for $x,\bar{x} \in S$ whenever $x+\Gamma(x,\bar{x})y \in S$. Then the sequence $\{x_n\}$ defined by (2.5), (3.1) converges to a solution x of the equation $Px = 0$. All x_n lie in S and the error estimate (2.14) holds.

Proof. It follows from the contractor inequality (2.13) with $x = x_n$, $\bar{x} = \bar{x}_n$ and $y = -Px_n$ that

$$(3.2) \qquad \|Px_{n+1}\| \leq BK(B+C)\,\|Px_n\|^2 \leq at_n^2 ,$$

where $t_0 = \eta$, $t_{n+1} = at_n^2$, $n = 0,1,\cdots$. The last inequality is easily verified by induction. Using induction again we arrive at (2.16) and we continue the proof in the same way as that of Theorem 2.2.

We assume now that the first divided difference operator F for P exists and is nonsingular. Then we consider the following iterative procedure investigated by Bel'tyukov [1] in a special case where $C_n = C$ and $0 < C \leq 1$.

$$(3.3) \qquad x_{n+1} = x_n - F(x_n,\bar{x}_n)^{-1}Px_n ,$$

for $n = 0,1,\cdots$, where $\bar{x}_n$ is defined by (3.1). Suppose in addition that condition (2.2) or (2.2*) is satisfied on $S = S(x_0,r) \subset D$. Under these hypotheses it follows from Lemma 2.1 that $\Gamma(x,\bar{x}) = F(x,\bar{x})^{-1}$ is a two point contractor for P. Thus, we can apply Theorem 3.1 to the procedure defined by (3.3), (3.1). In this way, we obtain the following theorem.

Theorem 3.2. Let $P:D \subset X \to Y$ be a closed nonlinear operator with domain D containing the ball $S = S(x_0,r)$. Suppose that there exists the first divided difference operator F for P satisfying (2.2) or (2.2*) on S. The operator F is nonsingular and satisfies condition (2.20). The initial element x_0 is chosen so as to satisfy

$$\|Px_0\| \leq \eta , \quad a = BK(B+C) , \quad q = a\eta < 1 , \quad r = B\eta t^* \quad \text{and} \quad t^* = \sum_{n=0}^{\infty} q^{2^n-1} .$$

Then the sequence $\{x_n\}$ defined by (3.3) and (3.1) converges to a solution x of the equation $Px = 0$. All x_n lie in S and the error estimate (2.14) holds.

<u>Proof</u>. By virtue of Lemma 2.1, $\Gamma(x,\bar{x}) = F(x,\bar{x})^{-1}$ is a bounded two point contractor satisfying the contractor inequality (2.13). It follows from (2.13) with $x = x_n$, $\bar{x} = \bar{x}_n$ and $y = -Px_n$ that using (3.1) we obtain (3.2) and continue as in the proof of Theorem 3.1.

<u>Remark 3.1</u>. Bel'tyukov [1] considered this method under restriction $0 < C < 1$. He uses a different argument and his arithmetic is based on that developed by Kantorovitch for the general Newton method. Our argument seems to be more general and rather simpler. His hypotheses are more restrictive. However, he does not make use of assumption (2.20). But this condition does hold in his case. This fact can be easily verified. Notice that condition (2.20) is essential for our argument but it also simplifies the reasoning.

It should be emphasized that both methods: the method considered by Laasonen and that by Bel'tyukov are different in form and were treated by different arguments. Nevertheless, it turns out that these methods can be handled from a common standpoint by using the same arithmetic. This fact is due to the unified approach given here which is based on the general contractor concept.

4. <u>Two point contractors with more general majorant functions</u>

So far we discussed only procedures involving bounded two point contractors. In this section, we consider a more general case. In this case, the contractor is bounded by a function. We introduce also a more general majorant function.

Let $P:D \subset X \to Y$ be a nonlinear closed operator and let $x_0 \in D$ be chosen so that $S = S(x_0,r) \subset D$, where the radius r will be defined below. Furthermore, let $Q(s,\bar{s},t,\bar{t}) \geq 0$, $s,\bar{s},t,\bar{t} \geq 0$, be a continuous function which is nondecreasing in each variable.

Let $\Gamma: D \times D \to L(Y \to X)$ be a two point contractor satisfying the following contractor inequality.

$$(4.1) \quad \|P(x+\Gamma(x,\bar{x})y) - Px - y\| \leq Q(\|\Gamma(x,\bar{x})y\|, \|x+\Gamma(x,\bar{x})y-\bar{x}\|, \|x-x_0\|, \|\bar{x}-\bar{x}_0\|)$$

for $x,\bar{x} \in S$ whenever $x+\Gamma(x,\bar{x})y \in D$. Let $B(t,\bar{t}) \geq 0, t,\bar{t} \geq 0$ be a function which is nondecreasing in each variable. We assume that the following estimate holds for the contractor

$$(4.2) \qquad \|\Gamma(x,\bar{x})\| \leq B(\|x-x_0\|, \|\bar{x}-\bar{x}_0\|)$$

for $x,\bar{x} \in S$, $x_0 \in S$ being fixed.

Now consider the general procedure

$$(4.3) \qquad x_{n+1} = x_n - \Gamma(x_n,\bar{x}_n)Px_n , \quad n = 0,1,\cdots ,$$

where $\bar{x}_n$ are chosen so as to satisfy the following inequalities

$$(4.4) \qquad \|\bar{x}_n-\bar{x}_0\| \leq \gamma_2\|x_n-x_0\|$$

$$(4.5) \qquad \|x_{n+1}-\bar{x}_n\| \leq \gamma_1\|x_{n+1}-x_n\|$$

for $n = 0,1,\cdots$, γ_1, γ_2 being constant numbers. We consider also the sequence $\{t_n\}$ defined by the following iterative process

$$(4.6) \qquad \begin{aligned} &t_{n+1} = t_n + B(t_n,\gamma_2 t_n)Q(t_n-t_{n-1},\gamma_1(t_n-t_{n-1}),t_{n-1},\gamma_2 t_{n-1}) , \\ &n = 0,1,\cdots \end{aligned}$$

with $t_0 = 0$. We assume that

$$t_n \to t^* \quad \text{as} \quad n \to \infty .$$

<u>Theorem 4.1</u>. Suppose that $P:D \subset X \to Y$ is a closed nonlinear operator and $x_0 \in D$ is such that

$$(4.7) \qquad \|\Gamma(x_0,\bar{x}_0)Px_0\| \leq \eta = t_1 \quad \text{and} \quad r = t^* .$$

If P has a two point contractor Γ satisfying (4.1) and (4.2) where Q is continuous and nondecreasing in each variable with

$Q(0,0,t^*,\gamma_2 t^*) = 0$, then all x_n lie in S and the sequence $\{x_n\}$ defined by (4.3) - (4.5) converges to a solution of equation $Px = 0$ and the error estimate holds

$$\|x-x_n\| \leq t^*-t_n \,, \quad n = 0,1,\cdots \,. \tag{4.8}$$

Proof. Since $Q(s,\bar{s},t,\bar{t}) \geq 0$ and $B(s,t) \geq 0$ for $s,\bar{s},t,\bar{t} \geq 0$, it follows from (4.6) that $0 = t_0 \leq t_1 \leq \cdots \leq t^*$. We have

$$\|x_1-x_0\| \leq t_1 \leq t^* \,, \quad \text{by (4.7) and (4.3)} \,.$$

Using (4.1) with $x = x_{n-1}$, $\bar{x} = \bar{x}_{n-1}$ and $y = -Px_{n-1}$ we show by induction that

$$\begin{aligned} \|Px_n\| &\leq Q(\|x_n-x_{n-1}\|,\|\bar{x}_n-\bar{x}_{n-1}\|,\|x_{n-1}-x_0\|,\|\bar{x}_{n-1}-\bar{x}_0\|) \\ &\leq Q(\|x_n-x_{n-1}\|,\gamma_1\|x_n-x_{n-1}\|,\|x_{n-1}-x_0\|,\gamma_2\|x_{n-1}-x_0\|) \\ &\leq Q(t_n-t_{n-1},\gamma_1(t_n-t_{n-1}),t_{n-1},\gamma_2 t_{n-1}) \,. \end{aligned} \tag{4.9}$$

$$\|x_n-x_0\| \leq \sum_{i=1}^{n} \|x_i-x_{i-1}\| \leq \sum_{i=1}^{n} (t_i-t_{i-1}) = t_n \leq t^* \,, \tag{4.10}$$

$$\begin{aligned} &\|x_{n+1}-x_n\| \leq \\ &\leq B(\|x_n-x_0\|,\|\bar{x}_n-\bar{x}_0\|)Q(\|x_n-x_{n-1}\|,\|\bar{x}_n-\bar{x}_{n-1}\|,\|x_{n-1}-x_0\|,\|\bar{x}_{n-1}-\bar{x}_0\|) \\ &\leq B(\|x_n-x_0\|,\gamma_2\|x_n-x_0\|)Q(\|x_n-x_{n-1}\|\gamma_1\|x_n-x_{n-1}\|,\|x_{n-1}-x_0\|,\gamma_2\|x_{n-1}-x_0\|) \\ &\leq B(t_n,\gamma_2 t_n)Q(t_n-t_{n-1},\gamma_1(t_n-t_{n-1}),t_{n-1},\gamma_2 t_{n-1}) = t_{n+1}-t_n \end{aligned} \tag{4.11}$$

by virtue of (4.3), (4.2) and (4.6). It follows from (4.10) that $x_n \in S$. By virtue of (4.11), we have

$$\|x_{n+p}-x_n\| \leq t_{n+p}-t_n \,. \tag{4.12}$$

Hence, the sequence $\{x_n\}$ converges to some element x and $Px_n \to 0$ as $n \to \infty$, by (4.9). Since P is closed, $Px = 0$. The error estimate (4.8) results from (4.12) by letting $p \to \infty$ in (4.12).

In our further discussion of general iterative procedures (4.3) we make use of some lemmas concerning Newton-type iterative methods in one dimensional space. These methods are examined for real valued functions considered as majorants for contractors (see Chapter 2, Section 4).

<u>Theorem 4.2</u>. Suppose that $P:D \subset X \to Y$ is a closed nonlinear operator and $x_0 \in D$ is such that satisfies (4.7). Furthermore, let Γ be a two point bounded contractor satisfying (4.1) and (4.2). Finally, assume that there exist functions $u(t)$ and $v(t)$ satisfying the hypotheses of Lemma 4.1 or 4.2, Chapter 2, and such that

$$(4.13)\quad B(s,\gamma_2 s)Q(s-t,\gamma_1(s-t),t,\gamma_2 t) \le [1/v(s)][u(s)-u(t)+v(t)(s-t)]$$

for $0 \le t < s \le t^*$, where t^* is the smallest root of the equation $u(t) = 0$; and

$$(4.14)\qquad r = t^*,\quad Q(0,0,t^*,\gamma_2 t^*) = 0 .$$

Then all $x_n \in S$ and the sequence $\{x_n\}$ defined by (4.3) - (4.5) converges to a solution of $Px = 0$. The error estimate (4.8) holds, where the sequence $\{t_n\}$ is defined by (4.10) or (4.11), Chapter 2, provided that $\|\Gamma(x_0,\bar{x}_0)Px_0\| \le \eta$.

<u>Proof</u>. Using the contractor inequality (4.1) with $x = x_{n-1}$, $\bar{x} = \bar{x}_{n-1}$ and $y = -Px_{n-1}$ we prove by induction the inequalities (4.9) - (4.12) exactly in the same way as in the proof of Theorem 4.1. By virtue of (4.13), we obtain

$$\begin{aligned}
\|x_{n+1}-x_n\| &= \|\Gamma(x_n,\bar{x}_n)Px_n\| \le \|\Gamma(x_n,\bar{x}_n)\|\cdot\|Px_n\| \le \\
&\le B(\|x_n-x_0\|,\|\bar{x}_n-\bar{x}_0\|)Q(\|x_n-x_{n-1}\|,\|\bar{x}_n-\bar{x}_{n-1}\|,\|x_{n-1}-x_0\|,\|\bar{x}_{n-1}-\bar{x}_0\|) \\
&\le B(\|x_n-x_0\|,\gamma_2\|x_n-x_0\|)Q(\|x_n-x_{n-1}\|,\gamma_1\|x_n-x_{n-1}\|,\|x_{n-1}-x_0\|,\gamma_2\|x_{n-1}-x_0\|) \\
&\le B(t_n,\gamma_2 t_n)Q(t_n-t_{n-1}),\gamma_1(t_n-t_{n-1}),t_{n-1},\gamma_2 t_{n-1}) \le \\
&\le [1/v(t_n)][u(t_n)-u(t_{n-1})+v(t_{n-1})(t_n-t_{n-1}) = t_{n+1}-t_n .
\end{aligned}$$

Hence we have

$$(4.15)\qquad \|x_{n+1}-x_n\| \le t_{n+1} - t_n$$

and

$$\|x_{n+1}-x_0\| \leq t_{n+1} < t^* = r \, , \quad n = 0,1,\cdots \, . \tag{4.16}$$

By Lemma 4.1 or 4.2, Chapter 2, the sequence $\{t_n\}$ defined there is increasing and converges toward the smallest root t^* of equation $u(t) = 0$. By virtue of (4.15), the sequence $\{x_n\}$ defined by (4.3) - (4.5) converges to some element x . It follows from (4.16) that $x_n \in S$ and $Px_n \to 0$ as $n \to \infty$ by virtue of (4.9) and (4.14). Since P is closed, we conclude that $Px = 0$. The error estimate (4.8) follows from (4.15).

As a special case of Theorem 4.2 we obtain the following

Theorem 4.3. In addition to the hypotheses of Theorem 4.2, assume instead of (4.13) that

$$B(s,\gamma_2 s)Q(s-t,\gamma_1(s-t),t,\gamma_2 t) \leq [1/(1-p_4 s)][p_1(s-t) + (p_1+p_3 t)](s-t) \tag{4.17}$$

for $0 \leq t < s \leq t^*$ and $u(t) = p_1 t^2 - (1-p_2)t + \eta$ and $v(t) = 1 - p_4 t$ satisfy the conditions of Lemma 4.3, Chapter 2. Then all assertions of Theorem 4.2 hold provided that the sequence $\{t_n\}$ and t^* are defined as in Lemma 4.3, Chapter 2.

Proof. The proof is exactly the same as that of Theorem 4.2, where (4.13) is replaced by (4.17).

As a special case of Theorem 4.3, we obtain the following

Theorem 4.4. Under the hypotheses of Theorem 4.2, instead of (4.13) assume that

$$Q(s-t,\gamma_1(s-t),t_1\gamma_2 t) \leq [(1/2)K(s-t)+d_0+d_1 t](s-t) \tag{4.18}$$

with d_0 , $d_1 \geq 0$ and in (4.2)

$$B(s,\gamma_2 s) \leq B/(1-BMs) \tag{4.19}$$

with $Bd_0 < 1$. Set t^* as in Lemma 4.3, Chapter 2; then all assertions of Theorem 4.2 hold provided that the sequence $\{t_n\}$ is defined as in that Lemma with $p_1 = (1/2)BK$, $p_2 = Bd_0$, $p_3 = Bd_1$, and $p_4 = BM$.

Proof. It follows from inequalities (4.18) and (4.19) that condition (4.17) is satisfied. Thus, the hypotheses of Theorem 4.3 are satisfied.

5. Approximate divided difference operators

Let us discuss some special cases of Theorem 4.4. We assume now that the first divided difference operator F for P exists and is nonsingular on S , that is $F(x,\bar{x})^{-1}$ exists for $x,\bar{x} \in S = S(x_0,r) \subset D$.

Let $A:D \times D \subset X \times X \to L(X \to Y)$. Suppose that $A(x_0,\bar{x}_0)$ is nonsingular and

$$\|F(x,\bar{x}) - A(x,\bar{x})\| \leq d_0 + \bar{d}_1[\|x-x_0\| + \|\bar{x} - \bar{x}_0\|] \tag{5.1}$$

for $x,\bar{x} \in S$. Consider the iterative procedure

$$x_{n+1} = x_n - A(x_n,\bar{x}_n)^{-1}Px_n , \quad n = 0,1,\cdots , \tag{5.2}$$

where $\bar{x}_n$ satisfy (4.4) and (4.5) or

$$\bar{x}_n = x_n + CPx_n , \quad n = 0,1,\cdots , \tag{5.3}$$

C being a constant number.

Theorem 5.1. Let $P:D \subset X \to Y$ be a closed nonlinear operator with domain D containing the ball $S = S(x_0,r)$, where x_0 is such that

$$\|A(x_0,\bar{x}_0)^{-1}Px_0\| \leq \eta , \|A(x_0,\bar{x}_0)^{-1}\| \leq B \text{ and } Bd_0 < 1 . \tag{5.4}$$

Suppose that the first divided difference operator F for P satisfies (2.2*) with K replaced by $\bar{K}$ and (5.1) for $x,\bar{x} \in S$, where A is such that

$$\|A(x,\bar{x}) - A(x_0,\bar{x}_0)\| \leq \bar{M}[\|x-x_0\| + \|\bar{x}-\bar{x}_0\|]$$

for $x,\bar{x} \in S$. Furthermore, assume that

$$\|I + CA(x,\bar{x})\| \leq \gamma_1 \quad \text{for } x,\bar{x} \in S , \tag{5.5}$$

(5.6) $$\|I + CF(x,\bar{x})\| \le \gamma_2 \quad \text{for} \quad x,\bar{x} \in S ,$$

where I is the identity operator. Let

$$p_1 = \frac{1}{2}\cdot BK , \quad p_2 = Bd_0 , \quad p_3 = Bd_1 \quad \text{and} \quad p_4 = BM ,$$

where

$$d_1 = \bar{d}_1(1+\gamma_2) , \quad K = 2\bar{K}\gamma_1 \quad \text{and} \quad M = \bar{M}(1+\gamma_2) .$$

Set t^* as in Lemma 4.3, Chapter 2, and $r = t^*$. Then the sequence $\{x_n\}$ defined by (5.2) and (5.3) remains in S and converges to a solution x of $Px = 0$.

Proof. We have for $\bar{x} = x + CPx$,

$$\|\bar{x}-\bar{x}_0\| = \|(x-x_0) + C(Px-Px_0)\| = \|(x-x_0) + CF(x,x_0)(x-x_0)\| \le$$
$$\le \|I + CF(x,x_0)\|\cdot\|x-x_0\| \le \gamma_2\|x-x_0\| , \quad \text{for} \quad x \in S ,$$

by (5.6). Hence, if $x = x_n$ and $\bar{x} = x_n$, then condition (4.4) holds. Furthermore, we have, by virtue of (5.5),

$$\|x+\Gamma(x,\bar{x})y-\bar{x}\| = \|\Gamma(x,\bar{x})y - CPx\| = \|\Gamma(x,\bar{x})y - CA(x,\bar{x})\Gamma(x,\bar{x})Px\| \le$$
$$\le \|I + CA(x,x)\|\cdot\|\Gamma(x,\bar{x})y\| \le \gamma_1\|\Gamma(x,\bar{x})y\|$$

for $y = -Px$. Hence, if $x = x_n$, $\bar{x} = x_n$ and $y = -Px_n$, then condition (4.5) holds.

For $x \in S(x_0,t^*)$, we have

$$\|A(x,\bar{x}) - A(x_0,\bar{x}_0)\| \le \bar{M}[\|x-x_0\| + \|\bar{x}-\bar{x}_0\|] \le \bar{M}(1+\gamma_2)\|x-x_0\| =$$
$$= M\|x-x_0\| < Mt^* \le M(2\bar{p}_1)^{-1} \le M(2\bar{p}_1-p_3)^{-1} \le Mp_4^{-1} = B^{-1} .$$

Hence, $A(x,\bar{x})$ is nonsingular and $\|\Gamma(x,\bar{x})\| \le B/(1-BM\|x-x_0\|)$, $x \in S(x_0,t^*)$; where $\Gamma(x,\bar{x}) = A(x,\bar{x})^{-1}$. Hence, condition (4.19) follows. Furthermore, we have

$$\|P(x+\Gamma(x,\bar{x})y) - Px - y\| \le \| P(x+\Gamma(x,\bar{x})y) - Px - F(x,\bar{x})\Gamma(x,\bar{x})y\| +$$
$$+ \|[F(x,\bar{x}) - A(x,\bar{x})]\Gamma(x,\bar{x})y\| \le \bar{K}\|\Gamma(x,\bar{x})y\|\cdot\|x+\Gamma(x,\bar{x})y-\bar{x}\| +$$

$$+ \{d_0 + \bar{d}_1(\|x-x_0\| + \|\bar{x}-\bar{x}_0\|)\}\|\Gamma(x,\bar{x})y\| \leq$$

$$\leq \bar{K}\gamma_1\|\Gamma(x,\bar{x})y\|^2 + \{d_0 + \bar{d}_1(1+\gamma_2)\|x-x_0\|\}\|\Gamma(x,\bar{x})y\| .$$

Hence, we obtain

$$\|P(x+\Gamma(x,\bar{x})y) - Px - y\| \leq (1/2)K\|\Gamma(x,\bar{x})y\|^2 + \{d_0+d_1\|x-x_0\|\}\|\Gamma(x,\bar{x})y\|$$

for $y = -Px$ and $x \in S$ whenever $x + \Gamma(x,\bar{x})y \in D$, where $d_1 = \bar{d}_1(1+\gamma_2)$. Therefore, Γ is a two point contractor with majorant function Q satisfying condition (4.18). Thus, we have shown that all hypotheses of Theorem 4.4 are satisfied and the proof is complete.

Using Lemma 4.4, Chapter 2, instead of Lemma 4.3, Chapter 2, we obtain the following theorem as a special case of Theorem 4.4.

Theorem 5.2. Let $P:D \subset X \to Y$ be a closed nonlinear operator with domain D containing the ball $S = S(x_0,r)$, where x_0 satisfies (5.4). Let Γ be a two point contractor satisfying the contractor inequality

$$\|P(x+\Gamma(x,\bar{x})y) - Px - y\| \leq K\|\Gamma(x,\bar{x})y\|\cdot\|x+\Gamma(x,\bar{x})y-\bar{x}\| \tag{5.7}$$

and

$$\|\Gamma(x,\bar{x})y\| \leq B/[1-M(\|x-x_0\| + \|\bar{x}-\bar{x}_0\|)] \tag{5.8}$$

for $x,\bar{x} \in S$ whenever $x + \Gamma(x,\bar{x})y \in S$. Put $a = B\max(2\gamma_1 K, M(1+\gamma_2))$ and assume that

$$h = a\eta \leq 1/2 \quad\text{and}\quad r = t^* = [1-(1-2h)^{1/2}]/a . \tag{5.9}$$

Then all x_n lie in S and the sequence $\{x_n\}$ defined by (4.3) - (4.5) converges to a solution x of equation $Px = 0$ and the error estimate holds

$$\|x-x_n\| \leq t^* - t_n \leq (a2^n)^{-1}(2h)^{2^n} , \tag{5.10}$$

for $n = 0,1,\cdots$, where the sequence $\{t_n\}$ is defined as in Lemma 4.4, Chapter 2.

Proof. We show by induction that

$$\|Px_n\| \leq K\gamma_1 \|\Gamma(x_{n-1},\bar{x}_{n-1})Px_{n-1}\|^2 = K\gamma_1 \|x_n - x_{n-1}\|^2$$

by virtue of (4.3), (4.5) and (5.7) with $x = x_{n-1}$, $\bar{x} = \bar{x}_{n-1}$ and $y = -Px_{n-1}$. It follows from (5.8) and (4.4) that

$$\|\Gamma(x_n,\bar{x}_n)\| \leq B/(1-M(1+\gamma_2)\|x_n - x_0\|) .$$

Hence, we obtain by induction

$$\|x_{n+1} - x_n\| \leq BK\gamma_1 \|x_n - x_{n-1}\|^2/(1-M(1+\gamma_2)\|x_n - x_0\|) \leq$$
$$\leq (1/2)a\|x_n - x_{n-1}\|^2/(1-a\|x_n - x_0\|) \leq (1/2)a(t_n - t_{n-1})^2/(1-at_n) = t_{n+1} - t_n$$

and

$$\|x_n - x_0\| \leq \sum_{i=1}^{n} \|x_i - x_{i-1}\| \leq \sum_{i=1}^{n} (t_i - t_{i-1}) = t_n \leq t^* ,$$

where the sequence $\{t_n\}$ is the same as that defined in Lemma 4.4, Chapter 2.

Now using this lemma we continue the same argument as in the proof of Theorem 4.4.

Replacing in Theorem 5.2 conditions (4.4) and (4.5) by (5.3), we obtain the following theorem.

Theorem 5.3. Under the hypotheses of Theorem 5.2 let the sequence $\{x_n\}$ be defined by (4.3) and (5.3). If, in addition,

$$\|I + CF(x,\bar{x})\| \leq \gamma \quad \text{for} \quad x,\bar{x} \in S ,$$

then all assertions of Theorem 5.2 hold.

Proof. It remains to prove that conditions (4.4) and (4.5) are satisfied. In fact, we have for $\bar{x} = x + CPx$,

$$\|\bar{x} - \bar{x}_0\| = \|(x-x_0) + C(Px - Px_0)\| = \|(x-x_0) + CF(x,x_0)(x-x_0)\| \leq$$
$$\leq \|I + CF(x,x_0)\| \cdot \|x - x_0\| \leq \gamma\|x - x_0\| .$$

Furthermore,

$$\|x+\Gamma(x,\bar{x})y-\bar{x}\| = \|\Gamma(x,\bar{x})y - CF(x,\bar{x})\Gamma(x,\bar{x})Px\| \leq \gamma\|\Gamma(x,\bar{x})y\|$$

for $y = -Px$.

Hence, putting $x = x_n$ and $\bar{x} = \bar{x}_n$, we conclude that (4.4) and (4.5) hold with $\gamma_1 = \gamma_2 = \gamma$.

Let us discuss a special case of Theorem 5.2 where the two point contractor Γ is replaced by the inverse of the first divided difference operator F . Thus, we consider the iterative procedure

$$x_{n+1} = x_n - F(x_n,\bar{x}_n)^{-1}Px_n \ , \tag{5.11}$$

where $\bar{x}_n$ are chosen so as to satisfy (4.4) and (4.5), $n = 0,1,\cdots$.

Theorem 5.4. Let $P:D \subset X \to Y$ be a closed nonlinear operator with domain D containing the ball $S = S(x_0,r)$. Suppose that the first divided difference operator F for P exists and satisfies (5.12) for $x,\bar{x},\xi,\bar{\xi} \in S$

$$\|F(x,\bar{x}) - F(\xi,\bar{\xi})\| \leq K(\|x-\xi\|+\|\bar{x}-\bar{\xi}\|) \ . \tag{5.12}$$

Assume that $F(x_0,\bar{x}_0)^{-1}$ exists and x_0 is such that

$$\|F(x_0,\bar{x}_0)^{-1}Px_0\| \leq \eta \quad \text{and} \quad \|F(x_0,\bar{x}_0)^{-1}\| \leq B \ . \tag{5.13}$$

Furthermore, put

$$\gamma = \max(2\gamma_1, 1+\gamma_2) \quad \text{and} \quad a = BK\gamma$$

and assume that

$$h = a\eta \leq 1/2 \quad \text{and} \quad r = t^* = [1-(1-2h)^{1/2}]/a \ . \tag{5.14}$$

Then all x_n lie in S and the sequence $\{x_n\}$ defined by (5.11), (4.4) and (4.5) converges to a solution x of equation $Px = 0$ and the error estimate (5.10) holds.

Proof. Let $\|\bar{x}-\bar{x}_0\| \leq \gamma_2\|x-x_0\|$. Then

$$\|F(x,\bar{x}) - F(x_0,\bar{x}_0)\| \le K(1+\gamma_2)\|x-x_0\| \le K\gamma\|x-x_0\| < B^{-1} ,$$

by virtue of (5.12) - (5.14), for $x \in S_1 = S(x_0,r)$ with $r = 1/a$. If $h < 1/2$, then $t^* < 1/a$ and $t^* = 1/a$ if $h = 1/2$. Thus, $F(x,\bar{x})$ has a nonsingular inverse for $x \in S(x_0,t^*)$ in particular for $x = x_n$ and $\bar{x} = \bar{x}_n$ satisfying (4.4). Now put $\Gamma(x,\bar{x}) = F(x,\bar{x})^{-1}$. Then we obtain

$$\|P(x+\Gamma(x,\bar{x})y) - Px - y\| = \|P(x+\Gamma(x,\bar{x})y) - Px - F(x,\bar{x})\Gamma(x,\bar{x})y\| =$$

$$= \|F(x+\Gamma(x,\bar{x})y,x)\Gamma(x,\bar{x})y - F(x,\bar{x})\Gamma(x,\bar{x})y\| \le$$

$$\le \|F(x+\Gamma(x,\bar{x})y,x) - F(x,\bar{x})\|\cdot\|\Gamma(x,\bar{x})y\| \le$$

$$\le K\|x+\Gamma(x,\bar{x})y-\bar{x}\|\cdot\|\Gamma(x,\bar{x})y\| .$$

Hence, it follows that condition (5.7) is satisfied. It results from (5.12) that condition (5.8) is also satisfied with $M = K$. Therefore, the hypotheses of Theorem 5.2 are satisfied and the proof is complete.

Consider now the iterative procedure defined by (5.11) and (5.3). The following theorem is a special case of Theorem 5.4.

Theorem 5.5. Under the hypotheses of Theorem 5.4 let the sequence $\{x_n\}$ be defined by (5.11) and (5.3). If, in addition,

$$\|I + CF(x,\bar{x})\| \le \gamma \quad \text{for} \quad x,\bar{x} \in S , \tag{5.15}$$

then all assertions of Theorem (5.4) hold.

Proof. It remains to prove that conditions (4.4) and (4.5) are fulfilled. In fact, this can be shown in the same way as in the proof of Theorem 5.3. Hence, it follows that all hypotheses of Theorem 5.4 are satisfied and the proof is complete.

Let us observe that setting in Theorem 5.5 $Px = \varphi(x) - x$ and $C = \mu$, where $0 < \mu \le 1$, we obtain the method investigated by Bel'tyukov [1]. In his case we have instead of (5.15):

$$\|I + CF(x,\bar{x})\| = \|I + \mu[\psi(x,\bar{x}) - I\| = \|(1-\mu)I + \mu\psi(x,\bar{x})\| \leq 1 - \mu + \alpha\mu = \gamma \ ,$$

where $\|\psi(x,\bar{x})\| \leq \alpha$ by assumption and $\psi(x,\bar{x})$ is the first divided difference operator for $\varphi: D \subset X \to X$, that is, $\psi: D \times D \subset X \times X \to L(X \to X)$. The argument used by Bel'tyukov is different and more restrictive. The argument used here is rather simpler and more general.

CHAPTER 8

CONTRACTORS AND ROOTS OF NONLINEAR FUNCTIONALS

Introduction. We have already shown that the contractor concept turned out to be an important tool in the development of a unified approach to the theory of solving operator equations by means of iterative procedures. Equations involving nonlinear functionals constitute a natural and important class of abstract equations. Now, it is our purpose to extend the contractor theory to iterative solutions of such equations. To serve this goal an independent concept of contractors for nonlinear functionals is introduced. In this way, a general theory can be developed for nonlinear functionals which is to some extent parallel to that for nonlinear operator. A new convergence proof for Newton's method for nonlinear functionals can be obtained as a special result of the general theory. Given a Banach space X let $F:D \subset X \to R$ (reals) be a nonlinear real valued functional and let $Q(s) \geq 0$ for $s \geq 0$. For simplicity we can assume that D is a closed ball $S = S(x_0, r)$ with center x_0 and radius r.

1. Contractors for nonlinear functionals

Definition 1.1. A mapping $\Gamma:S \subseteq X \to X$ is said to be a contractor for F with majorant function Q if the following inequality is satisfied.

$$(1.1) \qquad |F(x+t\Gamma(x)) - Fx - t| \leq Q(|t|)$$

for $x \in S$, whenever $x+t\Gamma(x) \in S$. Inequality (1.1) is called the contractor inequality. For practical purposes it suffices to require that (1.1) be satisfied for sufficiently small values of t depending on x and the function Q is supposed to be nondecreasing, continuous with $Q(0) = 0$.

Example. Let F be a nonlinear functional differentiable in the sense of Fréchet. Put $\Gamma(x) = [1/F'(x)h]h$, where $h \in X$ is such that $F'(x)h \neq 0$, $F'(x)$ being the Fréchet derivative of F at $x \in X$. Let $q < 1$ be an arbitrary positive number. Then the definition of $F'(x)$ yields

$$|F(x+\Gamma(x)) - Fx - t| = |F(x+t\Gamma(x)) - Fx - F'(x)[t\Gamma(x)]| \leq q|t|$$

for sufficiently small positive $|t|$ depending on x . Thus, the contractor inequality (1.1) is satisfied with $Q(t) = qt$.

A contractor Γ is said to be bounded on S if there is a positive constant B such that

$$\|\Gamma(x)\| \leq B \quad \text{for all} \quad x \in S . \tag{1.2}$$

A nonlinear functional F is said to be closed on S if $x_n \in S$, $x_n \to x$ and $Fx_n \to C$ imply $x \in S$ and $Fx = C$.

Our problem is to solve the equation

$$Fx = 0 , \quad x \in S , \tag{1.3}$$

where $F:S \subset X \to R$ is a nonlinear closed functional with a bounded contractor Γ . In this case we can assume that the contractor inequality has the following form

$$|F(x+t\Gamma(x)) - Fx - t| \leq at^2 + |bt| , \quad a > 0 , \quad b \geq 0 , \tag{1.4}$$

for $x \in S$, whenever $x+t\Gamma(x) \in S$.

For solving (1.3) we use the following iterative procedures

$$x_{n+1} = x_n - Fx_n\Gamma(x_n) , \tag{1.5}$$

$$t_{n+1} = Q(t_n) , \quad t_0 = \eta , \quad n = 0,1,\cdots , \tag{1.6}$$

where $Q(t) = at^2 + bt$.

<u>Theorem 1.1</u>. Suppose that $x_0 \in S$ is chosen so as to satisfy

$$S = S(x_0, Bt^*) \quad \text{and} \quad |Fx_0| \leq \eta \tag{1.7}$$

$$q = a\eta + b < 1 \ , \quad t^* = \eta/(1-q) \ . \tag{1.8}$$

If F has a bounded contractor Γ satisfying (1.4) and (1.2) for $x \in S$, then all x_n lie in S and the sequence $\{x_n\}$ defined by (1.5) converges to a solution x of equation (1.3) and the error estimate holds

$$\|x-x_n\| \leq B\eta \, q^n/(1-q) \ , \quad n = 0,1,\cdots \ . \tag{1.9}$$

<u>Proof</u>. It follows from (1.5) and (1.4) with $t = -Fx_n$ that

$$|Fx_{n+1}| \leq Q(|Fx_n|) \ , \quad n = 0,1,\cdots \ .$$

Furthermore, by virtue of (1.6) and (1.7), we prove by induction that

$$|Fx_n| \leq t_n \ , \quad n = 0,1,\cdots \ . \tag{1.10}$$

Now we prove also by induction that

$$t_{n+1} \leq qt_n \leq q^{n+1} \eta \leq \eta \ , \quad n = 0,1,\cdots \ . \tag{1.11}$$

In fact, we have, by (1.6) and (1.8),

$$t_1 = (at_0+b)t_0 = qt_0 \ .$$

Using the assumption for the induction we obtain

$$t_{n+1} = (at_n+b)t_n \leq (a\eta+b)t_n \leq qt_n \ .$$

Using a standard argument we derive from (1.5), (1.10) and (1.11) that if $m > n$, then

$$\|x_m-x_n\| \leq Bq^n\eta(1-q^{m-n-1})/(1-q) \ , \tag{1.12}$$

and all assertions of the theorem can be derived from (1.12) and using the fact that $Fx_n \to 0$ as $n \to \infty$, by virtue of (1.10), (1.11) and

(1.8).

<u>Remark 1.1</u>. A special case of Theorem 1.1 holds if $a = 0$ and $b = q < 1$, that is, the majorant function $Q(t) = qt$ is a linear function. The argument remains without change.

It should be mentioned that the iterative process (1.5) embraces a broad variety of different iteration procedures and the character of Theorem 1.1 is rather general. An application of this theorem can be obtained by using the following lemma.

<u>Lemma 1.1</u>. Let $F:D \subset X \to R$ be a nonlinear functional differentiable in the sense of Fréchet and let its derivative $F'(x)$ be Lipschitz continuous on some sphere $S = S(x_0,r) \subset D$, i.e., there exists a constant K such that

$$\|F'(x) - F'(y)\| \leq K\|x-y\| \tag{1.13}$$

for $x,y \in S$. Let $A:S \to X^*$ (the conjugate space) be a mapping such that

$$1/\|A(x)\| \leq B \quad \text{and} \quad \|F'(x) - A(x)\| \leq C \tag{1.14}$$

for $x \in S$. Let $0 < \beta < 1$ be any fixed number and put $\Gamma(x) = [1/A(x)h]h$, where $\|h\| = 1$ and $A(x)h \geq \beta\|A(x)\|$. Then Γ is a bounded contractor for F satisfying

$$\|\Gamma(x)\| \leq B/\beta \quad \text{for} \quad x \in S \tag{1.15}$$

and the contractor inequality

$$|F(x+\Gamma(x)) - Fx - t| \leq (1/2)K(B/\beta)^2t^2 + C(B/\beta)|t| \tag{1.16}$$

for $x \in S$ whenever $x+t\Gamma(x) \in S$.

<u>Proof</u>. We have

$$|F(x+t\Gamma(x)) - Fx - t| \leq |F(x+t\Gamma(x)) - Fx - F'(x)[t\Gamma(x)]| +$$

$$+ |F'(x)[t\Gamma(x)] - A(x)[t\Gamma(x)]| \leq (1/2)K\|\Gamma(x)\|^2t^2 + C\|\Gamma(x)\||t| \leq$$

$$\leq (1/2)K(B/\beta)^2 t^2 + C(B/\beta)|t|$$

by virtue of (1.13) - (1.16).

Theorem 1.2. Suppose that $F:D \subset X \to R$ satisfies the hypotheses of Lemma 1.1. If, in addition,

$$(1.17) \quad \begin{aligned} &|Fx_0| \leq \eta \ , \quad q = (1/2)(B/\beta)^2 K\eta + (B/\beta)C < 1 \ , \\ &r = (B/\beta)t^* \quad \text{and} \quad t^* = \eta/(1-q) \ , \end{aligned}$$

then all x_n lie in $S = S(x_0,r)$ and the sequence $\{x_n\}$ determined by (1.5) with $\Gamma(x)$ defined as in Lemma 1.1 converges to a solution x of equation (1.3) and the error estimate holds

$$(1.18) \quad \|x-x_n\| \leq (B/\beta)\eta q^n/(1-q) \ , \quad n = 0,1,\cdots \ .$$

Proof. By Lemma 1.1, F has a bounded contractor Γ satisfying the assumptions of Theorem 1.1 with $a = (1/2)(B/\beta)^2 K$ and $b = (B/\beta)C$. Since all hypotheses of Theorem 1.1 are fulfilled all assertions follow including the error estimate (1.9) in which B should be replaced by B/β and q is defined by (1.17).

Another application of Theorem 1.1 can be obtained by using the following lemma.

Lemma 1.2. Let $F:D \subset X \to R$ be a closed nonlinear functional and let $G:D \subset X \to R$ be a nonlinear functional differentiable in the sense of Fréchet on $S = S(x_0,r) \subset D$. Moreover, suppose that

$$\|G'(x) - G'(y)\| \leq K\|x-y\| \quad \text{and} \quad |(Fx-Gx) - (Fy-Gy)| \leq C\|x-y\|$$

for all $x,y \in S$ and

$$1/\|G'(x)\| \leq B \quad \text{for all} \quad x \in S \ .$$

Let $0 < \beta < 1$ be any fixed number and put $\Gamma(x) = [1/G'(x)h]h$, where $\|h\| = 1$ and $G'(x)h \geq \beta\|G'(x)\|$. Then Γ is a bounded contractor for F satisfying (1.15) and the contractor inequality (1.16).

Proof. We have

$$|F(x+t\Gamma(x)) - Fx - t| \leq |G(x+t\Gamma(x)) - Gx - G'(x)[t\Gamma(x)]| +$$

$$+ |[F(x+\Gamma(x)) - G(x+t\Gamma(x))] - [Fx-Gx]| \leq (1/2)K\|\Gamma(x)\|^2t^2 + C\|\Gamma(x)\||t| \leq$$

$$\leq (1/2)K(B/\beta)^2t^2 + C(B/\beta)|t| .$$

The following theorem gives an approximation method of finding a root of a nondifferentiable functional. The method yields sufficient conditions for existence of a root and convergence of the iterative procedure.

Theorem 1.3. Suppose that $F:D \subset X \to R$ satisfies the hypotheses of Lemma 1.2. If, in addition, F satisfies the conditions (1.17), then all x_n lie in S and the sequence $\{x_n\}$ determined by (1.5) with $\Gamma(x)$ defined as in Lemma 1.2 converges to a solution x of equation (1.3) and the error estimate (1.8) holds.

Proof. By Lemma 1.2, Γ is a bounded contractor for F satisfying condition (1.15) and the contractor inequality (1.16). Thus, all hypotheses of Theorem 1.1 are satisfied with $a = (1/2)(B/\beta)^2K$, $b = (B/\beta)C$ and q defined by (1.17).

2. Contractors bounded by functions

So far we discussed only those cases for which the contractor was bounded by a constant. In this section, we consider a more general case assuming that the contractor is bounded by a function.

Let $F:D \subset X \to R$ be a closed nonlinear functional and let $x_0 \in D$ be chosen so that $S = S(x_0,r) \subset D$, where r will be defined below. As before let $Q(t) \geq 0$ for $t \geq 0$. As far as applications are concerned, Q is supposed to be a nondecreasing continuous function with $Q(0) = 0$. Let $\Gamma:D \subset X \to X$ be a contractor for F with majorant function Q satisfying the contractor inequality

$$|F(x+t\Gamma(x)) - Fx - t| \leq Q(|t|\|\Gamma(x)\|) \tag{2.0}$$

for $x \in S$, whenever $x+t\Gamma(x) \in S$. We assume also that the following estimate holds for the contractor

$$\|\Gamma(x)\| \leq B(\|x-x_0\|) \quad \text{for} \quad x \in S, \tag{2.1}$$

where $B(t) \geq 0$ for $t \geq 0$ is a nondecreasing function.

Let us consider the sequence $\{t_n\}$ defined by

$$t_{n+1} = t_n + B(t_n)Q(t_n - t_{n-1}), \quad t_0 = 0, \quad n = 1,2,\cdots. \tag{2.2}$$

We assume that $t_n \to t^*$ as $n \to \infty$.

Theorem 2.1. Suppose that $F:D \subset X \to R$ is a closed nonlinear functional and $x_0 \in S \subset D$ is such that

$$|Fx_0| \cdot \|\Gamma(x_0)\| \leq t_1 \quad \text{and} \quad r = t^* = \lim_{n\to\infty} t_n. \tag{2.3}$$

If Γ is a contractor for F satisfying the contractor inequality (2.0) and (2.1), then all x_n lie in S and the sequence $\{x_n\}$ defined by (1.5) converges to a solution x of equation (1.3) and the error estimate holds

$$\|x-x_n\| \leq t^* - t_n, \quad n = 0,1,\cdots. \tag{2.4}$$

Proof. Since $Q(t) \geq 0$ and $B(t) \geq 0$ for $t \geq 0$, it follows from (2.2) that $0 = t_0 \leq t_1 \leq t_2 \leq \cdots t^*$. We have $\|x_1-x_0\| \leq t_1 \leq t^*$, by (2.3) and (1.5). By using the contractor inequality (1.3) with $x = x_{n-1}$, $t = -Fx_{n-1}$ we show by induction that

$$|Fx_n| \leq Q(\|Fx_{n-1}\Gamma(x_{n-1})\|) = Q(\|x_n-x_{n-1}\|) \leq Q(t_n-t_{n-1}), \tag{2.5}$$

$$\|x_n-x_0\| \leq \sum_{i=1}^{n} \|x_i-x_{i-1}\| \leq \sum_{i=1}^{n} (t_i-t_{i-1}) = t_n \leq t^*, \tag{2.6}$$

$$\|x_{n+1}-x_n\| \leq B(t_n)Q(t_n-t_{n-1}) = t_{n+1}-t_n, \tag{2.7}$$

by virtue of (1.5), (2.1) and (2.2). Hence,

$$\|x_{n+p}-x_n\| \leq t_{n+p}-t_n \quad \text{for} \quad n,p = 0,1,\cdots. \tag{2.8}$$

It follows from (2.6) that $x_n \in S$. By virtue of (2.8), the sequence $\{x_n\}$ converges to some element x and $Fx_n \to 0$ as $n \to \infty$, by (2.5). Since F is closed, $Fx = 0$. The error estimate (2.4) follows from (2.8) by letting $p \to \infty$ in (2.8).

Let us observe that the technique used in the theory of iterative methods for finding roots of nonlinear functionals turns out to be similar in a formal way to that for nonlinear operators. Moreover, the theory presented here appears to be a parallel development of the same for nonlinear operators. However, it should be emphasized that the theory of iterative methods for nonlinear functionals on a Banach space cannot be obtained as a particular case of the same theory for nonlinear operators although a functional is a particular case of an operator. As a classical example, let us consider the well-known Newton-Kantorovič method discussed in Chapters 1 and 2. It is bovious that in the formula (2.11), Chapter 1, for the Newton-Kantorovič method, the nonlinear operator P cannot be replaced by a nonlinear functional F. Nevertheless, a generalization of Newton's method for nonlinear functionals is possible and is given by formula (2.13), Chapter 1.

In order to discuss special cases of Theorem 2.1, we make use of some lemmas concerning Newton-type iterative methods in one dimensional space. These methods are examined for real valued functions considered as majorants for contractors. The same idea is exploited in Chapter 2 for approximate solutions of operator equations by using the contractor method with nonlinear majorant functions.

Let $Q(s,t) \geq 0$ for $s,t \geq 0$ be a continuous function which is nondecreasing in each variable. Introducing a more general majorant function Q of two independent variables, the contractor inequality (2.0) will be replaced by a more general one in such a way that Q becomes a function of $|t|\cdot\|\Gamma(x)\|$ and $\|x-x_0\|$, where x_0 is a given initial element of X. The situation in the case of nonlinear functionals appears to be formally similar to that for nonlinear operators (see Chapter 2, the inequality (5.1)). Thus, given a function Q let $F:D \subset X \to R$ be a nonlinear functional and let $\Gamma:D \subset X \to X$ be a contractor for F satisfying the following contractor inequality

$$(2.9) \qquad |F(x+t\Gamma(x)) - Fx - t| \leq Q(|t|\|\Gamma(x)\|, \|x-x_0\|),$$

for $x \in S = S(x_0,r) \subset D$ whenever $x+t\Gamma(x) \in D$.

Theorem 2.2. Suppose that $F:D \subset X \to R$ is a closed nonlinear functional and $S = S(x_0,r) \subset D$. Assume that Γ is a contractor for F satisfying (2.1) and the contractor inequality (2.9). Assume, in addition, that there exist functions $u(t)$ and $v(t)$ satisfying the hypotheses of Lemma 4.1 or 2.2, Chapter 2, and such that

$$(2.10) \qquad B(s)Q(s-t,t) \leq [1/v(s)][u(s) - u(t) + v(t)(s-t)]$$

for $0 \leq t < s \leq t^*$, where t^* is the smallest positive root of $u(t) = 0$. Finally, let

$$(2.11) \qquad |Fx_0|\|\Gamma(x_0)\| \leq \eta, \quad r = t^*, \quad Q(0,t^*) = 0.$$

Then all $x_n \in S$ and the sequence $\{x_n\}$ defined by (1.5) converges toward a solution of equation (1.3) and the error estimate (2.4) holds, where the sequence $\{t_n\}$ is defined as in Lemma 4.1, Chapter 2.

Proof. Using the contractor inequality (2.9) with $x = x_{n-1}$ and $t = -Fx_{n-1}$ we show by induction that

$$|Fx_n| \leq Q(|Fx_{n-1}|\|\Gamma(x_{n-1})\|, \|x_{n-1}-x_0\|) =$$
$$= Q(\|x_n-x_{n-1}\|, \|x_{n-1}-x_0\|) \leq Q(t_n-t_{n-1}, t_{n-1}).$$

Hence, by virtue of (2.1) and (1.5) we obtain

$$\|x_{n+1}-x_n\| = |Fx_n|\|\Gamma(x_n)\| \leq B(\|x_n-x_0\|)Q(\|x_n-x_{n-1}\|, \|x_{n-1}-x_0\|) \leq$$
$$\leq B(t_n)Q(t_n-t_{n-1}, t_{n-1}) \leq [1/v(t_n)][u(t_n) - u(t_{n-1}) +$$
$$+ v(t_{n-1})(t_n-t_{n-1})] = t_{n+1}-t_n,$$

by (2.10), (2.11).

Thus, we have

$$\|x_{n+1}-x_n\| \leq t_{n+1}-t_n \tag{2.12}$$

and

$$\|x_{n+1}-x_0\| \leq t_{n+1} < t^* = r \tag{2.13}$$

for $n = 0,1,\cdots$. By Lemma 4.1 or 4.2, Chapter 2, the sequence $\{t_n\}$ is increasing and converges toward the smallest root t^* of $u(t) = 0$. By virtue of (2.12), the sequence $\{x_n\}$ defined by (1.5) converges to some element x and we have $x_n \in S$, by (2.13). Since $Fx_n \to 0$ as $n \to \infty$ and F is closed, we conclude that $Fx = 0$. The error estimate (2.4) follows from (2.12) by using a standard argument.

Remark 2.1. If Q in (2.9) is a nondecreasing continuous function of one independent variable satisfying the contractor inequality (2.0) and $Q(0) = 0$, then condition (2.19) should be replaced by the following one

$$B(s)Q(s-t) \leq [1/v(s)][u(s) - u(t) + v(t)(s-t)]$$

for $0 \leq t < s \leq t^*$, where t^* is the smallest positive root of $u(t) = 0$. Then we obtain

$$|Fx_n| \leq Q(|Fx_{n-1}|\|\Gamma(x_{n-1})\|) = Q(\|x_n-x_{n-1}\|) \leq Q(t_n-t_{n-1})$$

and

$$\|x_{n+1}-x_n\| \leq B(\|x_n-x_0\|)Q(\|x_n-x_{n-1}\|) \leq B(t_n)Q(t_n-t_{n-1}) .$$

Everything else remains without change.

As a special case of Theorem 2.2, we obtain

Corollary 2.1. Under the hypotheses of Theorem 2.2 instead of (2.10), assume that

$$B(s)Q(w,t) \leq [1/(1-p_4 s)][p_1 w+p_2+p_3 t]w , \tag{2.14}$$

where $w = s-t$ and $0 \leq t < s \leq t^*$. Suppose that $\eta \leq (1-p_2)^2/4\bar{p}_1$ with $\bar{p}_1 = \max[p_1,(p_3+p_4)/2]$ and p_i , $i = 1,\cdots,4$, satisfy the conditions of Lemma 4.3, Chapter 2. Then all assertions of Theorem 2.2

hold true provided that $\{t_n\}$ and t^* are defined as in Lemma 4.3, Chapter 2.

Proof. It follows from (2.14) that

$$B(t_n)Q(t_n-t_{n-1},t_{n-1}) \leq$$

$$\leq [1/(1-p_4t_n)][\bar{p}_1(t_n-t_{n-1}) + p_2 + (2\bar{p}_1-p_4)t_{n-1}](t_n-t_{n-1}) = t_{n+1}-t_n .$$

Hence, the proof results from Theorem 2.2 and Lemma 4.3, Chapter 2.

As a special case of Corollary 2.1 we obtain

Corollary 2.2. Under the hypotheses of Theorem 2.2, set in (2.9)

$$Q(w,t) = (Kw+d_0+d_1t)w \tag{2.15}$$

with $d_0, d_1 \geq 0$ and in (2.1)

$$B(t) = B/(1-Ct) \tag{2.16}$$

with $Bd_0 < 1$. Let t^* and $\{t_n\}$ be defined as in Lemma 4.3, Chapter 2, with $p_1 = BK$, $p_2 = Bd_0$, $p_3 = Bd_1$ and $p_4 = C$. If $0 < \eta \leq (1-p_2)^2/4\bar{p}_1$ where $\bar{p}_1 = \max[BK,(Bd_1+C)/2]$, then all assertions of Theorem 2.2 hold true.

Proof. Conditions (2.15) and (2.16) imply (2.14). Hence, it follows that the hypotheses of Corollary 2.1 are satisfied.

A special case of Corollary 2.2 can be obtained by using the following lemma which yields another construction of a contractor for a nonlinear functional.

Lemma 2.1. Suppose that $F:D \subset X \to R$ is differentiable in the sense of Fréchet on the convex set $D_0 \subset D$ and

$$\|F'(x) - F'(y)\| \leq K\|x-y\| \quad \text{for } x,y \in D_0 .$$

Let $A:D_0 \to X^*$ be a mapping satisfying the following conditions

$$\|A(x)-A(x_0)\| \le M\|x-x_0\|$$

$$\|F'(x)-A(x)\| \le d_0 + d_1\|x-x_0\|$$

for all $x \in D_0$, where d_0, $d_1 \ge 0$ and $x_0 \in D$ is fixed. Assume, moreover, that $1/\|A(x_0)\| \le B$, and let $0 < \beta < 1$ be any fixed number. Put

$$\Gamma(x) = [1/A(x)h]h, \quad \text{where} \quad \|h\| = 1 \quad \text{and} \quad A(x)h \ge \beta\|A(x)\|, \quad h \in X.$$

Then Γ is a contractor for F satisfying the contractor inequality (2.9) with Q defined by (2.15) and

(2.17) $$\|\Gamma(x)\| \le \bar{B}/(1-\bar{B}M\|x-x_0\|)$$

for $\|x-x_0\| \le 1/\bar{B}M$, where $\bar{B} = B/\beta$.

<u>Proof</u>. We have

$$|F(x+t\Gamma(x)) - Fx - t| \le |F(x+t\Gamma(x)) - Fx - F'(x)[t\Gamma(x)]| +$$
$$|F'(x)[t\Gamma(x)] - A(x)[t\Gamma(x)]| \le (1/2)K\|\Gamma(x)\|^2t^2 +$$
$$+ (d_0+d_1\|x-x_0\|)\|t\Gamma(x)\| = Q(\|t\Gamma(x)\|, \|x-x_0\|).$$

Furthermore, we have

$$\|A(x)\| = \|A(x_0) + (A(x) - A(x_0))\| \ge \|A(x_0)\|(1-[1/\|A(x_0)\|]M\|x-x_0\|).$$

Hence, it follows that

$$\|\Gamma(x)\| \le 1/\beta\|A(x)\| \le \bar{B}/(1-BM\|x-x_0\|) \le \bar{B}/(1-\bar{B}M\|x-x_0\|)$$

if $\|x-x_0\| < 1/\bar{B}M$.

<u>Theorem 2.3</u>. Let $F:D \subset X \to R$ be a nonlinear functional satisfying the hypotheses of Lemma 2.1. Assume, in addition, that

$$|Fx_0|\cdot\|\Gamma(x_0)\| \le \eta, \quad \bar{B}d_0 < 1 \quad \text{and} \quad 0 < \eta \le (1-p_2)^2/4\bar{p}_1,$$

where $\bar{p}_1 = \max[\bar{B}K/2, B(d_1+M)/2]$. If $S(x_0,t^*) \subset D_0$, where t^* and $\{t_n\}$ are defined as in Lemma 4.3, Chapter 2, with $p_1 = \bar{B}K/2$,

$p_2 = \bar{B}d$, $p_3 = \bar{B}d_1$ and $p_4 = \bar{B}M$, then the sequence $\{x_n\}$ determined by (1.5) with Γ defined as in Lemma 2.1 remains in $S(x_0,t^*)$ and converges to a solution x of $Fx = 0$. The error estimate (2.4) holds true.

Proof. We conclude from Lemma 2.1 that assumptions (2.15) and (2.16) of Corollary 2.2 are satisfied with K replaced by $K/2$, B by $\bar{B}$ and C by $\bar{B}M$. Thus, the proof follows from Corollary 2.2.

Corollary 2.3. Suppose that $F:D \subset X \to R$ is a closed nonlinear functional and let $S = S(x_0,r) \subset D$. Assume that F has a contractor Γ satisfying the contractor inequality

$$|F(x+t\Gamma(x)) - Fx - t| \leq (1/2)Kt^2\|\Gamma(x)\|^2 \tag{2.18}$$

and

$$\|\Gamma(x)\| \leq B/(1-BK\|x-x_0\|) \tag{2.19}$$

for $x \in S$ whenever $x+t\Gamma(x) \in D$, where r , B , K and η are positive constants such that

$$|Fx_0|\|\Gamma(x_0)\| \leq \eta \quad \text{and} \quad h = BK\eta \leq 1/2 ,$$

and $r = t^* = [1-(1-2h)^{1/2}]/BK$. Then all x_n lie in S and the sequence $\{x_n\}$ defined by (1.5) converges toward a solution x of $Fx = 0$ and the error estimate holds.

$$\|x-x_n\| \leq t^*-t_n \leq (2h)^{2^n}/(BK2^n) , \quad \text{for} \quad n = 0,1,\cdots , \tag{2.20}$$

where $\{t_n\}$ is defined as in Lemma 4.4, Chapter 2.

Proof. Consider a particular case of Theorem 2.4 for which $d_0 = d_1 = 0$ in (2.15) and $M = K$ in (2.16). Then the hypotheses of Theorem 2.4 are satisfied with $Q(w,t) = Q(w) = (1/2)Kw^2$ and $B(t) = B/(1-BKt)$ and then Lemma 4.3 coincides with Lemma 4.4, Chapter 2, yielding the error estimate.

3. General iterative procedures with contractors

We continue the discussion of contractors bounded by functions. However, some change is made for the nonlinear majorant function Q. So let $F:D \subset X \to R$ be a nonlinear functional and let $\Gamma:D \subset X \to X$ be a contractor for F satisfying the following contractor inequality

$$(3.1) \qquad |F(x+t\Gamma(x)) - Fx - t| \leq Q(|t|)$$

for $x \in S = S(x_0,r) \subset D$ whenever $x + t\Gamma(x) \in D$.

In order to investigate the iterative procedure (1.5) for solving equation (1.3) we use the following two iterative sequences

$$(3.2) \qquad t_{n+1} = t_n + Q(t_n - t_{n-1}) \,, \quad n = 1,2,\cdots$$

with initial values $t_0 = 0$, $t_1 = Q(|Fx_0|)$

$$(3.3) \qquad s_{n+1} = s_n + B(s_n)(t_n - t_{n-1}) \,, \quad s_0 = 0 \,, \quad s_1 = |Fx_0|B(0) \,.$$

Then we obtain from (3.2) and (3.3)

$$(3.4) \qquad s_{n+1} = s_n + B(s_n)Q((s_n - s_{n-1})/B(s_{n-1})) \,, \quad n = 1,2,\cdots \,,$$

where $B(s)$ is the function involved in (2.1), $Q(t)$ being a non-decreasing continuous function with $Q(0) = 0$. We assume that both sequences $\{t_n\}$ and $\{s_n\}$ are convergent.

Theorem 3.1. Suppose that $F:D \subset X \to R$ is a closed nonlinear functional and $x_0 \in S \subset D$ is such that

$$(3.5) \qquad Q(|Fx_0|) = t_1 \,, \quad |Fx_0|B(0) = s_1 \quad \text{and} \quad r = s^* = \lim_{n\to\infty} s_n \,.$$

If Γ is a contractor for F satisfying the contractor inequality (3.1) and (2.1) and the sequences $\{t_n\}$, $\{s_n\}$ defined by (3.2) and (3.3), respectively, are convergent, then all x_n lie in S and the sequence $\{x_n\}$ defined by (1.5) converges to a solution x of equa tion (1.3) and the error estimate holds

$$(3.6) \qquad \|x - x_n\| \leq s^* - s_n \,, \quad n = 0,1,\cdots \,.$$

<u>Proof</u>. We show by induction using (3.1) with $x = x_n$ and $t = -Fx_n$ that

(3.7) $$|Fx_{n+1}| \leq Q(|Fx_n|) \leq Q(t_n - t_{n-1}) = t_{n+1} - t_n \,, \quad n = 1,2,\cdots \,,$$

by virtue of (3.2). Then we use also the induction to show that

$$\|x_{n+1} - x_n\| \leq B(s_n)(t_n - t_{n-1}) = s_{n+1} - s_n \,.$$

Hence it follows that

(3.8) $$\|x_n - x_0\| \leq s_n \,, \quad n = 0,1,\cdots \,,$$

by virtue of (1.5), (2.1) and (3.5). Hence, it follows by induction that

(3.9) $$\|x_n - x_m\| \leq s_n - s_m \,, \quad n > m = 0,1,\cdots \,.$$

Thus, the sequence $\{x_n\}$ converges to some x and we have $Fx_n \to 0$ as $n \to \infty$, by virtue of (3.7). We show by induction using (3.8) and (3.5) that $x_n \in S = S(x_0, r)$, $n = 1,2,\cdots$. The error estimate (3.6) follows from (3.9) by letting $n \to \infty$.

<u>Remark 3.1</u>. The convergence requirement for the sequence $\{t_n\}$ in Theorem 3.1 can be replaced by the following condition

(3.10) $$(s_{n+1} - s_n)/B(s_n) \to 0 \quad \text{as} \quad n \to \infty \,.$$

<u>Proof</u>. It remains to prove that $Fx_n \to 0$ as $n \to \infty$. In fact, we have by (3.7) and (3.3)

$$|Fx_{n+1}| \leq t_{n+1} - t_n = (s_{n+1} - s_n)/B(s_n) \quad \text{for} \quad n = 0,1,\cdots \,.$$

<u>Theorem 3.2</u>. Suppose that $F: D \subset X \to R$ is a closed nonlinear functional and $\Gamma: D \subset X \to X$ is a contractor for F satisfying (3.1) and (2.1) on $S = S(x_0, r) \subset D$ with both $B(t)$ and $Q(t)$ nondecreasing, Q being continuous and $Q(0) = 0$. In addition, suppose that there exist functions u and v satisfying the hypotheses of Lemma 4.1 or 4.2, Chapter 2, and such that

$$B(s)Q((s-t)/B(0)) \leq [1/v(s)][u(s)-u(t)+v(t)(s-t)] \tag{3.11}$$

for $0 \leq t \leq s \leq t^*$, where t^* is the smallest root of $u(t) = 0$.

Finally, let

$$|Fx_0|B(0) = s_1 \leq u(0)/v(0) < t^* = r . \tag{3.12}$$

Then all x_n lie in S and the sequence $\{x_n\}$ defined by (1.5) converges to a solution x of equation (1.3) and the error estimate holds

$$\|x-x_n\| \leq t^*-t_n , \quad n = 0,1,\cdots , \tag{3.13}$$

where the sequence $\{t_n\}$ and t^* are determined by Lemma 4.1 or 4.2, Chapter 2.

Proof. Using the same argument as in the proof of Theorem 3.1, we show by induction that

$$\|x_{n+1}-x_n\| \leq s_{n+1}-s_n , \quad n = 0,1,\cdots , \tag{3.14}$$

where the sequence $\{s_n\}$ is defined by (3.4). Using the induction again, we show that

$$s_{n+1}-s_n \leq t_{n+1}-t_n , \quad n = 0,1,\cdots . \tag{3.15$_n$}$$

In fact, assume that condition (3.15$_n$) is satisfied and $s_n \leq t_n$. Then we obtain $s_{n+1} \leq t_{n+1}$ and by virtue of (3.4), (2.11) and (3.11) with $s = t_{n+1}$, $t = t_n$

$$s_{n+2}-s_{n+1} = B(s_{n+1})Q((s_{n+1}-s_n)/B(s_n)) \leq B(t_{n+1})Q((t_{n+1}-t_n)/B(0)) \leq$$
$$\leq [1/v(t_{n+1})][u(t_{n+1})-u(t_n)+v(t_n)(t_{n+1}-t_n)] = t_{n+2}-t_{n+1} .$$

Thus, (3.15$_n$) and $s_n \leq t_n$ imply (3.15$_{n+1}$) and $s_{n+1} \leq t_{n+1}$. But $s_1-s_0 \leq t_1-t_0$, by (3.12), since $t_0 = s_0 = 0$. It follows from (3.14) and (3.15$_n$) that

$$\|x_n-x_m\| \leq s_n-s_m \leq t_n-t_m \quad \text{for } n > m = 0,1,\cdots . \tag{3.16}$$

By virtue of Lemma 4.1 or 4.2, Chapter 2, the sequence $\{t_n\}$ defined there with initial values $t_0 = 0$, $t_1 < t^*$ converges to t^* the smallest positive root of $u(t) = 0$. It follows then that the sequence $\{s_n\}$ is convergent, too, and condition (3.10) is also satisfied, since $B(t)$ is nondecreasing and $Q(t)$ is continuous with $Q(0) = 0$. By virtue of Theorem 3.1 and Remark 3.1, the sequence $\{x_n\}$ converges to a solution x of equation (1.3). The error estimate (3.13) results from (3.16) by letting $n \to \infty$.

4. Majorant functions of two variables

While the contractor remains still bounded by a function a further change for the nonlinear majorant function Q is made. Thus, let $F: D \subset X \to R$ be a nonlinear functional and let $\Gamma: D \subset X \to X$ be a contractor for F satisfying the following contractor inequality

$$|F(x+t\Gamma(x)) - Fx - t| \leq Q(|t|, \|x-x_0\|) \tag{4.1}$$

for $x \in S = S(x_0, r) \subset D$ whenever $x+t\Gamma(x) \in D$.

In order to investigate the iterative procedure (1.5) for solving equation (1.3) we consider the following two iterative numerical sequences

$$t_{n+1} = t_n + Q(t_n - t_{n-1}, s_n) , \quad n = 1,2,\cdots \tag{4.2}$$

with initial values $t_0 = 0$, $t_1 = Q(|Fx_0|, 0)$

$$s_{n+1} = s_n + B(s_n)(t_n - t_{n-1}) , \quad n = 1,2,\cdots \tag{4.3}$$

with initial values $s_0 = 0$, $s_1 = |Fx_0| B(0)$.

Hence we obtain from (4.2) and (4.3)

$$s_{n+1} = s_n + B(s_n) Q((s_n - s_{n-1})/B(s_{n-1}), s_{n-1}) , \quad n = 1,2,\cdots , \tag{4.4}$$

where $B(s)$ is the function in (2.1) and $Q(s,t) \geq 0$ for $s,t \geq 0$ is continuous nondecreasing in each variable with $Q(0,s^*) = 0$. We assume that both sequences $\{t_n\}$ and $\{s_n\}$ are convergent and $s^* = \lim_{n\to\infty} s_n$.

Theorem 4.1. Suppose that $F: D \subset X \to R$ is a closed nonlinear functional and $x_0 \in S \subset D$ is such that

(4.5) $$Q(|Fx_0|, 0) = t_1 \,, \quad |Fx_0| B(0) = s_1 \quad \text{and} \quad r = s^* \,.$$

If Γ is a contractor for F satisfying the contractor inequality (4.1) and (2.1) and the sequences $\{t_n\}$, $\{s_n\}$ defined by (4.2) and (4.3) respectively are convergent, then all x_n lie in S and the sequence $\{x_n\}$ defined by (1.5) converges to a solution x of equation (1.3) and the error estimate (3.6) holds true.

Proof. Using (4.1) with $x = x_n$ and $t = -Fx_n$ we show by induction that

(4.6) $$|Fx_{n+1}| \leq Q(|Fx_n|, \|x_n - x_0\|) \leq Q(t_n - t_{n-1}, s_n) = t_{n+1} - t_n \,, \quad n = 1,2,\cdots,$$

by virtue of (4.2) and

$$\|x_{n+1} - x_n\| \leq B(s_n)(t_n - t_{n-1}) = s_{n+1} - s_n \,.$$

Hence it follows that

(4.7) $$\|x_n - x_0\| \leq s_n \,, \quad n = 0,1,\cdots \,,$$

by virtue of (1.5), (2.1) and (4.5). Hence it follows by induction that

(4.8) $$\|x_n - x_m\| \leq s_n - s_m \,, \quad n > m = 0,1,\cdots \,.$$

Thus the sequence $\{x_n\}$ converges to some element x such that $Fx_n \to 0$ as $n \to \infty$, by virtue of (4.6). Using (4.5) and (4.7) we show by induction that $x_n \in S$ for $n = 1,2,\cdots$. The error estimate (3.6) with $\{s_n\}$ given by (4.3) follows from (4.8) by letting $n \to \infty$.

Remark 4.1. Remark 3.1 holds true in this case provided that $\{s_n\}$ is given by (4.3).

Theorem 4.2. Suppose that $F: D \subset X \to R$ is a closed nonlinear functional and $\Gamma: D \subset X \to X$ is a contractor for F satisfying (4.1) and (2.1) on

$S = S(x_0,r) \subset D$ with both $B(s)$ and $Q(s,t)$ nondecreasing in each variable, Q being continuous and $Q(0,s^*) = 0$. In addition, suppose that there exist functions u and v satisfying the hypotheses of Lemma 4.1 or 4.2, Chapter 2, and such that

$$B(s)Q((s-t)/B(0),t) \leq [1/v(s)][u(s)-u(t)+v(t)(s-t)] \tag{4.9}$$

for $0 \leq t < s \leq t^*$, where t^* is the smallest positive root of $u(t) = 0$. Finally, let

$$|Fx_0|B(0) = s_1 \leq u(0)/v(0) < t^* = r . \tag{4.10}$$

Then all x_n lie in S and the sequence $\{x_n\}$ defined by (1.5) converges to a solution x of equation (1.3) and the error estimate (3.13) holds.

Proof. The proof, mutatis mutandis, is the same as that of Theorem 3.2.

5. A new proof for Newton's method for nonlinear functionals

The Newton-Kantorovič method deals with nonlinear operators and is known to be a powerful method for solving nonlinear operator equations. A generalization of Newton's classical method for nonlinear functionals is given by Altman [18] and serves as a tool for solving functional equations involving nonlinear functionals. As a matter of fact, both methods are special cases of the general contractor method.

Let $F:D \subset X \to R$ be a nonlinear functional differentiable in the sense of Fréchet and let $S = S(x_0,r) \subset D$. Consider the following iterative process

$$x_{n+1} = x_n - [Fx_n/F'(x_n)h_n]h_n , \quad n = 0,1,\cdots , \tag{5.1}$$

where $h_n \in X$ is such that $\|h_n\| = 1$ and $F'(x_n)h_n \geq \beta\|F'(x_n)\|$, β being an arbitrary fixed number, $0 < \beta < 1$. The contractor method approach presented in this section yields a new convergence proof for the generalized Newton method (5.1). This result is the substance of the following theorem obtained as a special case of Corollary 2.3.

Theorem 5.1. Assume that $F:D \subset X \to R$ is differentiable in the sense of Fréchet in $S(x_0,r) \subset D$ and that $\|F'(x)-F'(y)\| \leq K\|x-y\|$ for all $x,y \in S(x_0,r)$. Suppose that x_0 is such that

$$|Fx_0|/F'(x_0)h_0 \leq \eta \quad \text{and} \quad 1/\|F'(x_0)\| \leq B$$

$$h = \bar{B}K\eta \leq 1/2 \quad \text{and} \quad r = t^* = [1-(1-2h)^{1/2}]/\bar{B}K ,$$

where $\bar{B} = B/\beta$.

Then all x_n lie in S and the sequence $\{x_n\}$ defined by (5.1) converges to a solution x of $Fx = 0$. Moreover, the error estimate hol holds

$$\|x-x_n\| \leq (2h)^{2^n}/(\bar{B}K2^n) , \quad \text{for} \quad n = 0,1,\cdots . \tag{5.2}$$

Proof. We have

$$|F(x+t\Gamma(x))-Fx-t| = |F(x+t\Gamma(x))-Fx-F'(x)[t\Gamma(x)]| \leq (1/2)K\|\Gamma(x)\|^2t^2,$$

where $\Gamma(x) = [1/F'(x)h]h$ with $\|h\| = 1$ and $F'(x)h \geq \beta\|F'(x)\|$.

Thus, $\Gamma(x)$ is a contractor satisfying the contractor inequality (2.18). Furthermore, we have

$$\|F'(x)\| = \|F'(x_0)+(F'(x)-F'(x_0))\| \geq \|F'(x_0)\|(1-[1/\|F'(x_0)\|]K\|x-x_0\|) .$$

Hence it follows that

$$\|\Gamma(x)\| \leq 1/\beta\|F'(x)\| \leq \bar{B}/(1-\bar{B}K\|x-x_0\|)$$

if $\|x-x_0\| < 1/\bar{B}M$. Thus, $\Gamma(x)$ satisfies the inequality (2.19) with B replaced by $\bar{B}$. The remaining hypotheses of Corollary 2.3 are also satisfied with B replaced by $\bar{B}$. The error estimate (5.2) follows from (2.20) with B replaced by $\bar{B}$ and the proof is complete.

Remark 5.1. Suppose that the Banach space X has the following property: if f is a linear continuous functional on X, then there exists an element $h \in X$ such that $\|h\| = 1$ and $fh = \|f\|$. For instance, every reflexive Banach space has this property. For Banach spaces with this property we can put $\beta = 1$ in all theorems involving β and in particular for the generalized Newton method (5.1) for nonlinear functionals.

CHAPTER 9

CONTRACTORS AND SECANT METHODS FOR FINDING ROOTS OF NONLINEAR FUNCTIONALS

Introduction. In this chapter, we apply the idea of two point secant methods to the problem of finding approximate roots of nonlinear functionals. For this purpose a further development of the contractor concept for nonlinear functionals in Banach spaces is presented. An analysis of the contractor notion applied to two point secant methods for nonlinear operators makes it clear how to generalize the contractor concept for nonlinear functionals. In this way, the notion of a two point contractor for a nonlinear functional is introduced. Let us recall that the contractor notion for a nonlinear functional has been introduced independently in a sense that the definition of a contractor for a nonlinear functional is not a simple specialization of the same definition of a contractor for a nonlinear operator. In the same way, the definition of a two point contractor for a nonlinear functional is being introduced independently. On the basis of the concept of two point contractors for nonlinear functionals various two point iteration methods can be introduced for finding roots of nonlinear functionals. By analogy, these methods are called two point secant methods for nonlinear functionals.

1. Two point contractors for nonlinear functionals

Given a Banach space X and a nonlinear functional $F:D \subset X \to R$ (reals) let $Q(s,t) \geq 0$ for $t,s \geq 0$. For simplicity we can assume that D is a closed ball $S = S(x_0,r)$ with center x_0 and radius r.

Definition 1.1. A mapping $\Gamma:D \times D \to X$ is said to be a two point contractor for F with majorant function Q if the following inequality is satisfied

(1.1) $$|F(x+t\Gamma(x,\bar{x})) - Fx - t| \leq Q(|t|, \|x-\bar{x}\|)$$

for $x, \bar{x} \in D$ and real t whenever $x + t\Gamma(x,\bar{x}) \in D$ - the domain of F. Inequality (1.1) is called the contractor inequality.

Let us recall that a contractor for F as defined in Chapter 8 is a mapping $\Gamma: D \to X$. Thus, the concept of a two point contractor for a nonlinear functional F generalizes the concept of a contractor for F. The following contractor inequality is a special case of (1.1).

$$(1.1^*) \qquad |F(x+t\Gamma(x,\bar{x})) - Fx - t| \leq 0(|t|) + K\|x-\bar{x}\| ,$$

where K is a constant and $0 \leq 0(s)/s \to 0$ as $s \to 0$, $s \geq 0$. It follows from (1.1^*) that $\tilde{\Gamma}(x) = \Gamma(x,x)$ is an inverse derivative for F as defined in Chapter 1, and, consequently, a contractor in the sense of Altman [1].

By means of analogy, the first divided difference functional of F is a linear bounded functional

$$G: D \times D \to X^* \quad \text{(the conjugate space)}$$

which satisfies the relationship

$$G(x,\bar{x})(x-\bar{x}) = Fx - F\bar{x} \quad \text{for} \quad x,\bar{x} \in D \subset X .$$

For each $x,\bar{x} \in D$, choose $y \in X$ such that $G(x,\bar{x})y \neq 0$ and put $\Gamma(x,\bar{x}) = y/G(x,\bar{x})y$.

Lemma 1.1. If $\|\Gamma(x,\bar{x})\| \leq B$ with some constant B and

$$(1.2) \qquad \|G(x,\bar{x}) - G(\xi,\bar{\xi})\| \leq K[\|x-\xi\| + \|\bar{x}-\bar{\xi}\|]$$

for $x,\bar{x},\xi,\bar{\xi} \in D$, then $\Gamma(x,\bar{x})$ is a two point contractor for F.

Proof. Put $h = t\Gamma(x,\bar{x})$; then we obtain

$$|F(x+h) - Fx - t| = |F(x+h) - Fx - G(x,\bar{x})h|$$

$$= |G(x+h,x)h - G(x,\bar{x})h| \leq K[\|h\| + \|x-\bar{x}\|]\,\|h\|$$

$$\leq KB(B|t| + \|x-\bar{x}\|)|t| = Q(|t|, \|x-\bar{x}\|) .$$

Note that Lemma 1.1 remains true if one replaces condition (1.2) by the following one

$$(1.2^*) \qquad |Fx - F\bar{x} - G(\xi,\bar{x})(x-\bar{x})| \leq K\|x-\bar{x}\|\cdot\|x-\xi\| .$$

In fact, we have for $h = t\Gamma(x,\bar{x})$,

$$|F(x+h) - Fx - t| = |F(x+h) - Fx - G(x,\bar{x})h| \leq K\|h\|\cdot\|x+h-\bar{x}\| \leq$$

$$\leq K\|h\|(\|h\| + \|x-\bar{x}\|) \leq KB|t|(B|t| + \|x-\bar{x}\|) = Q(|t|,\|x-\bar{x}\|) .$$

A two point contractor is said to be bounded if there exists a constant B such that

$$(1.3) \qquad \|\Gamma(x,\bar{x})\| \leq B \quad \text{for} \quad x,\bar{x} \in D \subset X .$$

Consider the equation

$$(1.4) \qquad Fx = 0 ,$$

where $F:D \to R$ is a closed nonlinear functional, i.e., $x_n \in D$, $x_n \to x$ and $Fx_n \to c$ imply $x \in D$ and $Fx = c$.

We assume that F has a two point contractor Γ satisfying the contractor inequality (1.1) with $Q(s,t)$ continuous and nondecreasing in each variable and $Q(0,0) = 0$.

In order to solve (1.4) we consider the following general iterative procedure

$$(1.5) \qquad x_{n+1} = x_n - Fx_n\Gamma(x_n,\bar{x}_n) , \quad n = 0,1,\cdots ,$$

where $\bar{x}_n$ will be defined below. We consider also the iterative numerical sequence $\{t_n\}$ defined as follows

$$(1.6) \qquad t_{n+1} = t_n + Q(t_n-t_{n-1},C(t_n-t_{n-1})) , \quad n = 0,1,\cdots$$

with initial values $t_0 = 0$, $t_1 = Q(\eta,C\eta)$, the constant C being defined below. We investigate the iterative procedure (1.5) under the assumption that the elements $\bar{x}_n$ are chosen so as to satisfy

$$(1.7) \qquad \|x_n-\bar{x}_n\| \leq C|Fx_n| , \quad n = 0,1,\cdots .$$

Assume also that the sequence $\{t_n\}$ determined by (1.6) has a limit t^*.

<u>Theorem 1.1</u>. Let $F:D \to R$ be a closed nonlinear functional with domain D containing the ball $S = S(x_0,r)$, where x_0 is such that $|Fx_0| \leq \eta$. Suppose that $\Gamma: S \times S \to X$ is a two point contractor for F satisfying (1.3) and the contractor inequality (1.1) for $x,\bar{x} \in S$ and real t whenever $x+t\Gamma(x,\bar{x}) \in D$. Let $\bar{x}_n$ in (1.5) be chosen so as to satisfy (1.7). Assume also that the sequence $\{t_n\}$ defined by (1.5) has a limit t^* and $r = Bt^*$. Then the sequence $\{x_n\}$ defined by (1.5) and (1.7) converges to a solution x of equation (1.4). All x_n lie in S and the following error estimate holds

$$\|x-x_n\| \leq B(t^*-t_n), \quad n = 0,1,\cdots . \tag{1.8}$$

<u>Proof</u>. It follows from the contractor inequality (1.1) with $t = -Fx_n$, $x = x_n$, $\bar{x} = \bar{x}_n$ and by virtue of (1.5) using also (1.7) that

$$|Fx_{n+1}| \leq Q(|Fx_n|, C|Fx_n|), \quad n = 0,1,\cdots . \tag{1.9}$$

Now we show by induction that

$$|Fx_n| \leq t_n - t_{n-1} \quad \text{for} \quad n = 1,2,\cdots . \tag{1.10$_n$}$$

In fact, using (1.9) with $n = 0$ we obtain

$$|Fx_1| \leq Q(|Fx_0|, C|Fx_0|) \leq Q(\eta, C\eta) = t_1 - t_0 ,$$

by virtue of (1.6). Assuming (1.10$_n$) and using (1.9) again we conclude by virtue of (1.6) that

$$|Fx_{n+1}| \leq Q(t_n - t_{n-1}, C(t_n - t_{n-1})) = t_{n+1} - t_n$$

whence (1.10$_{n+1}$) follows.

Furthermore, (1.3), (1.5) and (1.10$_n$) imply

$$\|x_{n+m} - x_n\| \leq \sum_{i=n}^{n+m-1} \|x_{i+1} - x_i\| \leq B \sum_{i=n}^{n+m-1} |Fx_i|$$

$$\leq B \sum_{i=n}^{n+m-1} (t_{i+1} - t_i) = B(t_{n+m} - t_n) ,$$

that is,

$$\|x_{n+m}-x_n\| \leqq B(t_{n+m}-t_n) \quad (1.11)$$

Inequality (1.11) shows that $\{x_n\}$ is a Cauchy sequence and, consequently, has a limit x. Since $Fx_n \to 0$ as $n \to \infty$, by (1.10_n), and F is closed, we conclude that $Fx = 0$. Putting $n = 0$ in (1.11) we obtain

$$\|x_m-x_0\| \leqq Bt^* \quad \text{for} \quad m = 1,2,\cdots .$$

Letting $m \to \infty$ in (1.11), we obtain the error estimate (1.8) and the proof is complete.

Now let us discuss some special cases of condition (1.7) in conjunction with procedure (1.5).

Suppose that $\bar{x}_n$ in (1.5) is defined as follows

$$\bar{x}_{n+1} = x_{n+1} - Fx_{n+1}\Gamma(x_n,\bar{x}_n) \;, \quad n = 0,1,\cdots . \quad (1.12)$$

Assuming (1.3), we conclude from (1.12) that condition (1.7) is satisfied with constant $C = B$. Suppose now that the majorant function Q has the form $Q(s,t) = KB(Bs+t)s$, where B, K are some positive constants. Thus, the contractor inequality (1.1) becomes

$$|F(x+t\Gamma(x,\bar{x})) - Fx - t| \leqq BK(B|t| + \|x-\bar{x}\|)\,|t| . \quad (1.13)$$

In this case the procedure defined by (1.5) and (1.12) is quadratically convergent and we have the following.

<u>Theorem 1.2</u>. Let $F{:}D \to R$ be a closed nonlinear functional with domain D containing the ball $S = S(x_0,r)$, where $|Fx_0| \leqq \eta$, $\|x_0-\bar{x}_0\| \leqq B\eta$, $a = 2KB^2$, $q = a\eta < 1$, $r = B\eta t^*$ and $t^* = \Sigma_{n=0}^{\infty} q^{2n-1}$.

Suppose that Γ is a two point contractor for F satisfying (1.3) and the contractor inequality (1.13) for $x,\bar{x} \in S$ whenever $x+t\Gamma(x,\bar{x}) \in S$.

Then the sequence $\{x_n\}$ defined by (1.5) and (1.12) converges to a solution x of the equation $Fx = 0$. All x_n lie in S and the error estimate holds:

$$\|x - x_n\| \leq B\eta t^* q^{2^n - 1} . \tag{1.14}$$

Proof. It follows from the contractor inequality (1.13) with $x = x_n$, $\bar{x} = \bar{x}_n$ and $t = -Fx_n$ that

$$|Fx_{n+1}| \leq 2KB^2 |Fx_n|^2 \leq at_n^2 = t_{n+1} \tag{1.15}$$

where $t_0 = \eta$, $n = 1,2,\cdots$ and $|Fx_1| \leq a\eta^2 = t_1$. We prove by induction that

$$t_n = \eta q^{2^n - 1} \quad \text{where} \quad t_{n+1} = at_n^2 , \quad n = 0,1,\cdots . \tag{1.16}$$

Hence, by virtue of (1.5), (1.3), (1.15) and (1.16) we obtain

$$\|x_{n+1} - x_n\| \leq B\eta q^{2^n - 1}$$

$$\|x_{n+1} - x_0\| \leq B\eta \sum_{i=0}^{n} q^{2^i - 1} < B\eta t^* .$$

The latter inequality shows that all x_n lie in S. Further, we have

$$\|x_{n+m} - x_n\| \leq B\eta \sum_{i=n}^{n+m-1} q^{2^i - 1} < B\eta t^* q^{2^n - 1} . \tag{1.17}$$

Thus, $\{x_n\}$ is a Cauchy sequence and has a limit x. It follows from (1.15), (1.16) and $q < 1$ that $Fx_n \to 0$ as $n \to \infty$. Since F is closed, we conclude that $Fx = 0$. The error estimate (1.14) follows from (1.17) by letting $m \to \infty$.

We assume now that the first divided difference functional G for F exists and satisfies condition (1.2) or (1.2*) on $S = S(x_0, r) \subset D$. Under these hypotheses, Lemma 1.1 shows that $\Gamma(x,\bar{x}) = y/G(x,\bar{x})y$, as a mapping from $D \times D$ into X, where $y \in X$ is such that $G(x,\bar{x})y \neq 0$, is a bounded two point contractor for F provided that

$$\|\Gamma(x,\bar{x})\| \leq B \quad \text{for} \quad x,\bar{x} \in S \tag{1.18}$$

holds true. Let us consider the following iterative procedure.

$$x_{n+1} = x_n - Fx_n \Gamma(x_n, \bar{x}_n) \tag{1.19}$$

$$\bar{x}_{n+1} = x_{n+1} - Fx_{n+1}\Gamma(x_n,\bar{x}_n) \tag{1.20}$$

for $n = 0,1,\cdots$.

Applying Theorem 1.2 to this procedure we obtain

Theorem 1.3. Let $F:D \to R$ be a closed nonlinear functional with domain D containing the ball S . Suppose that there exists the first divided difference functional G for F satisfying (1.2) or (1.2*) and (1.18). Let the initial elements x_0 and $\bar{x}_0$ be chosen so as to satisfy $|Fx_0| \leq \eta$, $\|x_0-\bar{x}_0\| \leqq B\eta$, $r = B\eta t^*$, $a = 2KB^2$, $q = a\eta < 1$ and $t^* = \Sigma_{n=0}^{\infty} q^{2^n-1}$. Then the sequence $\{x_n\}$ defined by (1.19) and (1.20) converges to a solution x of the equation $Fx = 0$. All x_n lie in S and the error estimate (1.14) holds.

Proof. By virtue of Lemma 1.1, Γ is a two point bounded contractor for F satisfying the contractor inequality (1.13). It is easy to see that all hypotheses of Theorem 1.2 are satisfied. Hence, the proof follows immediately.

Remark 1.1. It is sufficient to assume instead of (1.18) that

$$\|\Gamma(x_n,\bar{x}_n\| = \|y_n\|/|G(x_n,\bar{x}_n)y_n| \leqq B \quad \text{for} \quad n = 0,1,\cdots .$$

2. Two point contractors and consistent approximations

Consistent approximations were introduced by Ortega (see Ortega and Rheinboldt ([2], NR 11.2-3. page 363) as a more general notion than a divided difference operator of a nonlinear operator. The case discussed here is related to nonlinear functionals. It turns out that the notion of consistent approximations is a particular case of the two point contractor concept for nonlinear functionals.

Let $F:D \to R$ be a nonlinear functional with Fréchet derivative $F(x)$, $x \in D \subset X$.

Definition 2.1. A mapping $J:D \times D \to X^*$ is called a strongly consistent approximation to F' on $S = S(x_0,r) \subset D$ if there exists a constant c such that

$$(2.1) \qquad \|F'(x) - J(x,\bar{x})\| \leq c\|x-\bar{x}\| \quad \text{for} \quad x,\bar{x} \in S \ .$$

Lemma 2.1. Let $F:D \to R$ be continuously differentiable in S and let the Fréchet derivative F' be Lipschitz continuous on S with Lipschitz constant $\bar{K}$. Suppose that J is a strongly consistent approximation to F' satisfying (2.1). For each $x,\bar{x} \in S$ choose $y \in X$ such that $J(x,\bar{x})y \neq 0$. Then

$$(2.2) \qquad \Gamma(x,\bar{x}) = y/J(x,\bar{x})y \ , \quad x,\bar{x} \in S$$

is a two point bounded contractor for F provided that Γ satisfies the condition

$$(2.3) \qquad \|\Gamma(x,\bar{x})\| \leq B \quad \text{for} \quad x,\bar{x} \in S \ ,$$

where B is some constant.

Proof. For $h = t\Gamma(x,\bar{x})$ we have

$$|F(x+t\Gamma(x,\bar{x})) - Fx - t| \leq |F(x+h) - Fx - F'(x)h| + |F'(x)h - J(x,\bar{x})h| \leq$$
$$\leq \frac{1}{2}\bar{K}\|h\|^2 + c\|x-\bar{x}\|\cdot\|h\| \leq \frac{1}{2}\bar{K}B^2t^2 + cB\|x-\bar{x}\|\cdot|t|$$

by virtue of (2.1) and (2.3). Hence, we obtain the following contractor inequality

$$(2.4) \qquad |F(x+t\Gamma(x,\bar{x})) - Fx - t| \leq BK(B|t| + \|x-\bar{x}\|)|t|$$

for $x,\bar{x} \in S$ whenever $x+t\Gamma(x,\bar{x}) \in S$, where $K = \max(\frac{1}{2}\bar{K},c)$.

Theorem 2.1. Let $F:D \to R$ be a nonlinear continuously differentiable functional with domain D containing S and let F' be Lipschitz continuous on S with Lipschitz constant $\bar{K}$. Let J be a strongly consistent approximation to F' satisfying (2.1) and (2.3). Suppose that x_0, $\bar{x}_0$ are chosen so as to satisfy $|Fx_0| \leq \eta$, $\|x_0-\bar{x}_0\| \leq B\eta$, $a = 2KB^2$, $q = a\eta < 1$, $r = B\eta t^*$ and $t^* = \Sigma_{n=0}^{\infty} q^{2n-1}$ where K $K = \max(\frac{1}{2}\bar{K},c)$. Then the sequence $\{x_n\}$ defined by (1.5) and (1.12) with Γ determined by (2.2) converges to a solution x of $Fx = 0$.

All x_n lie in S and the error estimate (1.14) holds true.

Proof. Since the hypotheses of Lemma 2.1 are satisfied, it follows that Γ defined by (2.2) is a two point bounded contractor for F satisfying the contractor inequality (2.4). It is easily seen that all hypotheses of Theorem 1.2 are satisfied and the proof follows immediately.

Remark 2.1. Let us observe that condition (2.3) can be replaced by the following one

$$\|\Gamma(x_n,\bar{x}_n)\| = \|y_n\|/|J(x_n,\bar{x}_n)y_n| \leqq B \quad \text{for} \quad n = 0,1,\cdots .$$

Remark 2.2. Let $0 < \beta < 1$ be any fixed number. One can choose y in (2.2) so as to satisfy the following conditions

$$J(x,\bar{x})y \geq \beta\|J(x,\bar{x})\|, \|y\| = 1 . \tag{2.5}$$

Now assume that there exists a constant $\bar{B}$ such that

$$1/\|J(x,\bar{x})\| \leqq \bar{B} \quad \text{for} \quad x,\bar{x} \in S . \tag{2.6}$$

Then we obtain by virtue of (2.5), (2.6)

$$\|\Gamma(x,\bar{x})\| \leqq B/\beta \quad \text{for} \quad x,\bar{x} \in S . \tag{2.7}$$

Thus, in all cases condition (2.3) can be replaced by (2.7).

In some Banach spaces one can put $\beta = 1$. This is the case, for instance, in reflexive Banach spaces.

3. Two point contractors bounded by functions

In this section, we discuss iterative procedures involving two point contractors bounded by functions. We also introduce a more general majorant function.

Let $F:D \to R$ be a nonlinear closed functional with domain D containing the ball $S = S(x_0,r)$ with radius r to be defined below. Furthermore, let $Q(s,\bar{s},t,\bar{t}) \geq 0$ for $s,\bar{s},t,\bar{t} \geq 0$ be a continuous function which is nondecreasing in each variable.

Let $\Gamma: D \times D \to X$ be a two point contractor for F satisfying the following contractor inequality

$$|F(x+t\Gamma(x,\bar{x})) - Fx - t| \leqq Q(\|t\Gamma(x,\bar{x})\|, \|x+t\Gamma(x,\bar{x}) - \bar{x}\|, \|x-x_0\|, \|\bar{x}-\bar{x}_0) \tag{3.1}$$

for $x,\bar{x} \in S$ whenever $x+t\Gamma(x,\bar{x}) \in S$.

Let $B(t,\bar{t}) \geq 0$ for $t,\bar{t} \geq 0$ be a function which is nondecreasing in each variable. We assume that the following estimate holds true for the contractor Γ

$$\|\Gamma(x,\bar{x})\| \leqq B(\|x-x_0\|, \|\bar{x}-\bar{x}_0\|) \tag{3.2}$$

for $x,\bar{x} \in S$ with $x_0 \in S$ being fixed.

Now consider the general two point contractor procedure

$$x_{n+1} = x_n - Fx_n\Gamma(x_n,\bar{x}_n) , \quad n = 0,1,\cdots , \tag{3.3}$$

where $\bar{x}_n$ are chosen so as to satisfy the following inequalities

$$\|x_{n+1}-\bar{x}\| \leqq \gamma_1\|x_{n+1}-x_n\| \tag{3.4}$$

$$\|\bar{x}_n-\bar{x}_0\| \leqq \gamma_2\|x_n-x_0\| \tag{3.5}$$

for $n = 0,1,\cdots; \gamma_1$ and γ_2 being constant numbers. We also consider the scalar sequence $\{t_n\}$ defined by the following iterative process

$$t_{n+1} = t_n + B(t_n,\gamma_2 t_n)Q(t_n-t_{n-1},\gamma_1(t_n-t_{n-1}),t_{n-1},\gamma_2 t_{n-1}) , \tag{3.6}$$

$n = 0,1,\cdots$ with initial value $t_0 = 0$.

We assume that

$$t_n \to t^* \quad \text{as} \quad n \to \infty .$$

<u>Theorem 3.1</u>. Suppose that $F: D \to R$ is a closed nonlinear functional with domain D containing the sphere S and x_0 is such that

$$\|Fx_0\Gamma(x_0,\bar{x}_0)\| \leqq \eta = t_1 \quad \text{and} \quad r = t^* . \tag{3.7}$$

If Γ is a two point contractor for F satisfying (3.1) and (3.2) where Q is continuous and nondecreasing in each variable with $Q(0,0,t^*,\gamma_2 t^*) = 0$, then all x_n lie in S and the sequence $\{x_n\}$ defined by (3.3) - (3.5) converges to a solution x of equation $Fx = 0$ and the error estimate holds

$$\|x-x_n\| \leq t^*-t_n\,, \quad n = 0,1,\cdots . \tag{3.8}$$

Proof. Since $Q(s,\bar{s},t,\bar{t}) \geq 0$ for $s,\bar{s},t,\bar{t} \geq 0$, it follows from (3.6) that $0 = t_0 \leq t_1 \leq t_2 \leq \cdots \leq t^*$. We have

$$\|x_1-x_0\| \leq t_1 \leq t^*$$

by (3.7) and (3.3). We show by induction using the contractor inequality (3.1) with $x = x_{n-1}$, $\bar{x} = \bar{x}_{n-1}$ and $t = -Fx_{n-1}$ that

$$\begin{aligned} |Fx_n| &\leq Q(\|x_n-x_{n-1}\|,\|x_n-\bar{x}_{n-1}\|,\|x_n-x_0\|,\|\bar{x}_{n-1}-x_0\|) \\ &\leq Q(\|x_n-x_{n-1}\|,\gamma_1\|x_n-x_{n-1}\|,\|x_n-x_0\|,\gamma_2\|x_{n-1}-x_0\|) \\ &\leq Q(t_n-t_{n-1},\gamma_1(t_n-t_{n-1}),t_{n-1},\gamma_2 t_{n-1}) \end{aligned} \tag{3.9}$$

$$\|x_n-x_0\| \leq \sum_{i=1}^{n}\|x_i-x_{i-1}\| \leq \sum_{i=1}^{n}(t_i-t_{i-1}) = t_n \leq t^*\,, \tag{3.10}$$

$$\begin{aligned} &\|x_{n+1}-x_n\| \\ &\leq B(\|x_n-x_0\|,\|\bar{x}_n-\bar{x}_0\|)Q(\|x_n-x_{n-1}\|,\|\bar{x}_n-\bar{x}_{n-1}\|,\|x_{n-1}-x_0\|,\|\bar{x}_{n-1}-\bar{x}_0\|) \\ &\leq B(\|x_n-x_0\|,\gamma_2\|x_n-x_0\|)Q(\|x_n-x_{n-1}\|,\gamma_1\|x_n-x_{n-1}\|,\|x_{n-1}-x_0\|,\gamma_2\|x_{n-1}-x_0\|) \\ &\leq B(t_n,\gamma_2 t_n)Q(t_n-t_{n-1},\gamma_1(t_n-t_{n-1}),t_{n-1},\gamma_2 t_{n-1}) = t_{n+1}-t_n\,, \end{aligned} \tag{3.11}$$

by virtue of (3.3), (3.2) and (3.6). It follows from (3.10) that $x_n \in S$. We conclude from (3.11) that

$$\|x_{n+p}-x_n\| \leq t_{n+p}-t_n \quad \text{for} \quad n,p = 1,2,\cdots . \tag{3.12}$$

Hence, the sequence $\{x_n\}$ converges to some element x and we have $Fx_n \to 0$ as $n \to \infty$, by (3.9). Since F is closed, $Fx = 0$. The error estimate (3.8) results from (3.12) by letting $p \to \infty$ in (3.12).

Our further discussion of general iterative procedures (3.3) is based on certain lemmas concerning Newton type iterative methods in one dimensional space. These methods are used in order to establish the convergence of the sequences $\{t_n\}$ in (3.6) (see Chapter 2, Lemmas 4.1 - 4.4).

Theorem 3.2. Let $F:D \to R$ be a closed nonlinear functional with domain D containing S and let $\Gamma:D \times D \to X$ be a two point contractor satisfying (3.1) and (3.2). Assume that there exist two functions $u(t)$ and $v(t)$ satisfying the hypotheses of Lemma (4.1) or (4.2), Chapter 2, and

$$(3.13)\quad B(s,\gamma_2 s)Q(s-t,\gamma_1(s-t),t,\gamma_2 t) \leq [1/v(s)][u(s)-u(t)+v(t)(s-t)]$$

for $0 \leq t < s \leq t^*$, where t^* is the smallest positive root of $u(t) = 0$, and

$$(3.14)\quad Q(0,0,t^*,\gamma_2 t^*) = 0 .$$

Finally let $x_0 \in D$ be such that it satisfies (3.7). Then all x_n lie in S and the sequence $\{x_n\}$ defined by (3.3) - (3.5) converges to a solution x of $Fx = 0$. The error estimate (3.8) holds, where the sequence $\{t_n\}$ is defined as in Lemma 4.1, Chapter 2, provided that $\|Fx_0\Gamma(x_0,\bar{x}_0)\| \leq \eta$.

Proof. Using the contractor inequality (3.1) with $x = x_{n-1}$, $\bar{x} = \bar{x}_{n-1}$ and $t = -Fx_{n-1}$ we prove by induction the inequalities (3.9) - (3.12) exactly in the same way as in the proof of Theorem 3.1. By virtue of (3.13), we obtain

$$\|x_{n+1}-x_n\| = \|Fx_n\Gamma(x_n,\bar{x}_n\| \leq \|\Gamma(x_n,\bar{x}_n)\|\cdot|Fx_n|$$

$$\leq B(\|x_n-x_0\|,\|\bar{x}_n-\bar{x}_0\|)Q(\|x_n-x_{n-1}\|,\|x_n-\bar{x}_{n-1}\|,\|x_n-x_0\|,\|\bar{x}_n-\bar{x}_0\|)$$

$$\leq B(\|x_n-x_0\|,\gamma_2\|x_n-x_0\|)Q(\|x_n-x_{n-1}\|,\gamma_2\|x_n-x_{n-1}\|,\|x_n-x_0\|,\gamma_2\|\bar{x}_n-\bar{x}_0\|)$$

$$\leq B(t_n,\gamma_2 t_n)Q(t_n-t_{n-1},\gamma_1(t_n-t_{n-1}),t_{n-1},\gamma_2 t_{n-1})$$

$$\leq [1/v(t_n)][u(t_n)-u(t_{n-1})+v(t_{n-1})(t_n-t_{n-1})] = t_{n+1}-t_n .$$

Hence, we have

$$\|x_{n+1}-x_n\| \leqq t_{n+1}-t_n \tag{3.15}$$

and

$$\|x_{n+1}-x_0\| \leqq t_{n+1} < t^* = r\ , \quad n = 0,1,\cdots . \tag{3.16}$$

By Lemma 4.1 or 4.2, Chapter 2, the sequence $\{t_n\}$ defined there by (4.10) or (4.11) is increasing and converges toward the smallest positive root t^* of $u(t) = 0$. By virtue of (3.15), the sequence $\{x_n\}$ defined by (3.3) - (3.5) converges to some element x. It follows from (3.16) that x_n is in S and $Fx_n \to 0$ as $n \to \infty$ by virtue of (3.9) and (3.14). Since F is closed, we conclude that $Fx = 0$. The error estimate (3.8) follows from the inequality (3.15) by using a standard argument and the proof of the theorem is completed.

As a special case of Theorem 3.2, we obtain the following

Corollary 3.1. In addition to the hypotheses of Theorem 3.2 assume instead of (3.13) that

$$\begin{aligned} &B(s,\gamma_2 s)Q(s-t,\gamma_1(s-t),t,\gamma_2 t) \\ &\leqq [1/(1-p_4 s)][p_1(s-t)+p_2+p_3 t](s-t) \end{aligned} \tag{3.17}$$

for $0 \leqq t < s \leqq t^*$ and $u(t) = \bar{p}_1 t^2 - (1-p_2)t + \eta$ and $v(t) = 1 - p_4 t$ satisfy the conditions of Lemma 4.3, Chapter 2. Then all assertions of Theorem 3.2 hold provided the sequence $\{t_n\}$ and t^* are defined as in Lemma 4.3, Chapter 2.

Proof. The proof is exactly the same as that of Theorem 3.2.

As a special case of Corollary 3.1 we obtain the following

Corollary 3.2. Under the hypotheses of Corollary 3.1 instead of (3.13) assume that

$$Q(s-t,\gamma_1(s-t),t,\gamma_2 t) \leqq [\tfrac{1}{2}K(s-t)+d_0+d_1 t](s-t) \tag{3.18}$$

with $d_0, d_1 \geqq 0$ and

(3.19) $$B(s,\gamma_2 s) \leq B/(1-\alpha s)$$

in (3.2) with $Bd_0 < 1$ and $\eta \leq (1-Bd_0)^2/4\bar{p}_1$, where $\bar{p}_1 = \max[p_1,(p_3+p_4)/2]$.

Then all assertions of Theorem 3.2 hold provided the sequence $\{t_n\}$ and t^* are defined as in Lemma 4.3, Chapter 2, with $p_1 = \frac{1}{2}BK$, $p_2 = Bd_0$, $p_3 = Bd_1$ and $p_4 = \alpha$.

Proof. Since condition (3.17) is fulfilled, the hypotheses of Corollary 3.1 are satisfied.

4. Approximate first ordered divided differences

Let $F:D \to R$ be a nonlinear functional with domain D containing the ball $S = S(x_0,r)$. Assume that the first divided difference functional $G(x,\bar{x})$ of F exists for $x,\bar{x} \in S$ and satisfies

(4.1) $$|Fx - F\bar{x} - G(\xi,\bar{x})(x-\bar{x})| \leq \bar{K}\|x-\bar{x}\|\cdot\|x-\xi\|$$

for $x,\bar{x} \in S$. Let $J:D \times D \to X^*$ be a mapping such that

(4.2) $$\|G(x,\bar{x}) - J(x,\bar{x})\| \leq d_0 + d_1[\|x-x_0\| + \|\bar{x}-\bar{x}_0\|]$$

for $x,\bar{x} \in S$.

Lemma 4.1. For each $x,\bar{x} \in S$ let $y \in X$ be such that $J(x,\bar{x})y \neq 0$ and put $\Gamma(x,\bar{x}) = y/J(x,\bar{x})y$. Then $\Gamma:S \times S \to X$ is a two point contractor for F satisfying the following contractor inequality

(4.3) $$\begin{aligned}&|F(x+t\Gamma(x,\bar{x})) - Fx - t|\\ &\leq \bar{K}\|t\Gamma(x,\bar{x})\|\{\|x+t\Gamma(x,\bar{x})-\bar{x}\| + d_0 + d_1(\|x-x_0\| + \|\bar{x}-\bar{x}_0\|)\}\end{aligned}$$

for $x,\bar{x} \in S$ whenever $x+t\Gamma(x,\bar{x}) \in S$.

Proof. We have by virtue of (4.1) and (4.2)

$$\begin{aligned}|F(x+t\Gamma(x,\bar{x})) - Fx - t| &\leq |F(x+t\Gamma(x,\bar{x})) - Fx - G(\bar{x},x)t\Gamma(x,\bar{x})|\\ &+ |[G(x,\bar{x}) - J(x,\bar{x})]t\Gamma(x,\bar{x})| \leq \bar{K}|t\Gamma(x,\bar{x})\|\cdot\|x+t\Gamma(x,\bar{x})-\bar{x}\|\\ &+ \{d_0+d_1(\|x-x_0\| + \|x-x_0\|)\}\|t\Gamma(x,\bar{x})\| .\end{aligned}$$

Now let us discuss the iterative procedure (3.3) under condition (1.7) to be shown as a special case of condition (3.4).

Lemma 4.2. Suppose that there exists a constant number δ such that

(4.4) $$\|J(x,\bar{x})\| \leqq \delta \quad \text{for} \quad x,\bar{x} \in S .$$

Then condition (1.7) implies (3.4) with $\gamma_1 = 1+\delta C$.

Proof. It follows from (4.4) that

(4.5) $$1/\|\Gamma(x,\bar{x})\| = |J(x,\bar{x})y|/\|y\| \leqq \delta \quad \text{for} \quad x,\bar{x} \in S .$$

Hence, we obtain by virtue of (1.7) and (4.5)

(4.6) $$\|x_n-\bar{x}_n\| \leqq C|Fx_n| = C\|Fx_n\Gamma(x_n,\bar{x}_n)\|/\|\Gamma(x_n,\bar{x}_n)\| \leqq \delta C\|x_{n+1}-x_n\|$$

for $n = 0,1,\cdots$ provided $x_n \in S$. Since

$$\|x_{n+1}-\bar{x}_n\| = \|x_n-Fx_n\Gamma(x_n,\bar{x}_n) - \bar{x}_n\|$$
$$\leqq \|x_n-\bar{x}_n\| + \|Fx_n\Gamma(x_n,\bar{x}_n)\| \leqq (1+\delta C)\|x_{n+1}-x_n\| ,$$

by virtue of (4.6), we conclude that condition (3.4) with $\gamma_1 = 1+\delta C$ results from (1.7) and (4.4).

Suppose now that there exists a constant $\bar{M}$ such that

(4.7) $$\|J(x,\bar{x}) - J(x_0,\bar{x}) \leqq \bar{M}[\|x-x_0\| + \|\bar{x}-\bar{x}_0\|] \quad \text{for} \quad x,\bar{x} \in S .$$

Let $\beta < 1$ be any positive fixed number and for each pair $x,\bar{x} \in S$ choose $y \in X$ so as to satisfy

(4.8) $$J(x,\bar{x})y \geqq \beta\|J(x,\bar{x})\| \quad \text{and} \quad \|y\| = 1 .$$

Then we obtain

$$\|J(x,\bar{x})\| \geqq \|J(x_0,\bar{x}_0)\|\,|1 - [\|J(x_0,\bar{x}_0) - J(x,\bar{x})\|]/\|J(x_0,\bar{x}_0)\|\,|$$
$$\geqq \|J(x_0,\bar{x}_0)\|(1 - \bar{B}\bar{M}[\|x-x_0\| + \|\bar{x}-\bar{x}_0\|]) \geqq (1 - BM\|x-x_0\|)/B$$

provided $\|\bar{x}-\bar{x}_0\| \leqq \gamma_2\|x-x_0\|$,

(4.9) $$1/\|J(x_0,\bar{x}_0)\| \leqq \bar{B} ,$$

and $B = \bar{B}/\beta$, $M = \bar{M}(1+\gamma_2)$, $BMr < 1$. Then we obtain by virtue of (4.8) and (4.9)

(4.10) $$\|\Gamma(x,\bar{x})\| \leqq B/(1 - BM\|x-x_0\|) \quad \text{for} \quad x,\bar{x} \in S \quad \text{if} \quad BMr \leqq 1 .$$

Theorem 4.1. Let $F:D \to R$ be a closed nonlinear functional with domain D containing S . Assume that the first divided difference functional of F exists and satisfies (4.1), (4.2) and (4.4). Let x_0 , $\bar{x}_0$ be chosen so as to satisfy (4.9) and

$$\|Fx_0\Gamma(x_0,\bar{x}_0)\| \leqq \eta \quad \text{and} \quad Bd_0 < 1 .$$

Let

$$\eta \leqq (1 - Bd_0)^2/4\bar{p}_1 , \quad K = 2\bar{K}(1+\delta C) , \quad M = \bar{M}(1+\gamma_2) , \quad p_4 = BM$$

and set $r = t^*$ where t^* is defined as in Lemma 4.3, Chapter 2, with

$$p_1 = \frac{1}{2}BK, \quad p_2 = Bd_0 , \quad p_3 = Bd_1 , \quad d_1 = \bar{d}_1(1+\gamma_2) .$$

Then the sequence $\{x_n\}$ defined by (3.3) under condition (1.7) remains in S and converges to a solution x of $Fx = 0$.

Proof. It follows from Lemma 4.1 that Γ is a two point contractor for F satisfying the contractor inequality (4.3). Since $Mt^* \leqq 1/B$, Γ satisfies (4.10). It results from (4.3) and (4.6) that

$$|F(x+t\Gamma(x,\bar{x})) - Fx - t| \leqq \bar{K}(1+\delta C)\|t\Gamma(x,\bar{x})\|^2$$
$$+ \{d_0+\bar{d}_1(1+\gamma_2)\|x-x_0\|\}\|t\Gamma(x,\bar{x})\|$$

for $t = -Fx$, $\|x-\bar{x}\| \leqq C\|Fx\|, \|\bar{x}-\bar{x}_0\| \leqq \gamma_2\|x-x_0\|$. Hence, we obtain that condition (3.18) is satisfied with $\gamma_1 = 1 + \delta C$, $K = 2\bar{K}\gamma_1$ and $d_1 = \bar{d}_1(1+\gamma_2)$. Thus, the hypotheses of Corollary 3.2 are satisfied and the proof is complete.

5. Quadratic convergence

A special case of Theorem 3.1 will be discussed. This case yields quadratic convergence and is based on Lemma 4.4, Chapter 2.

<u>Theorem 5.1</u>. Let $F:D \to R$ be a closed nonlinear functional with domain D containing the ball $S = S(x_0,r)$. Let $\Gamma:D \times D \to X$ be a two point contractor for F satisfying the contractor inequality

$$|F(x+t\Gamma(x,\bar{x})) - Fx - t| \leq K\|t\Gamma(x,\bar{x})\| \cdot \|x+t\Gamma(x,\bar{x}) - \bar{x}\| \tag{5.1}$$

and

$$\|\Gamma(x,\bar{x})\| \leq B/[1 - M(\|x-x_0\| + \|\bar{x}-\bar{x}_0\|)] \tag{5.2}$$

for $x,\bar{x} \in S$ whenever $x+t\Gamma(x,\bar{x}) \in S$. Put $a = \max[2B\gamma_1 K, M(1+\gamma_2)]$ and assume that

$$h = a\eta \leq \frac{1}{2} \quad \text{and} \quad r = t^* = [1-(1-2h)^{1/2}]/a. \tag{5.3}$$

Let $x_0, \bar{x}_0 \in D$ be chosen so as to satisfy

$$\|Fx_0\Gamma(x_0,\bar{x}_0\| \leq \eta. \tag{5.4}$$

Then all x_n lie in S and the sequence $\{x_n\}$ defined by (3.3) - (3.5) converges to a solution x of $Fx = 0$ and the error estimate holds:

$$\|x-x_n\| \leq t^*-t_n \leq (a2^n)^{-1}(2h)^{2n}, \tag{5.5}$$

for $n = 0,1,\cdots$, where the sequence $\{t_n\}$ is defined as in Lemma 4.4, Chapter 2.

<u>Proof</u>. We show by induction that

$$|Fx_n| \leq K\gamma_1\|Fx_{n-1}\Gamma(x_{n-1},\bar{x}_{n-1})\|^2 = K\gamma_1\|x_n-x_{n-1}\|^2$$

by virtue of (3.3), (3.4) and (5.1) with $x = x_{n-1}$, $\bar{x} = \bar{x}_{n-1}$ and $t = -Fx_{n-1}$. It follows from (5.2) and (3.5) that

$$\|\Gamma(x_n,\bar{x})\| \leq B/(1 - M(1+\gamma_2)\|x-x_0\|).$$

Hence, we obtain by induction

$$\|x_{n+1}-x_n\| \leq BK\gamma_1\|x_n-x_{n-1}\|^2/(1 - M(1+\gamma_2)\|x_n-x_0\|)$$

$$\leq \frac{1}{2}a\|x_n-x_{n-1}\|^2/(1-a\|x_n-x_0\|) \leq \frac{1}{2}a(t_n-t_{n-1})^2/(1-at_n) = t_{n+1}-t_n$$

and

$$\|x_n - x_0\| \leq \sum_{i=1}^{n} \|x_i - x_{i-1}\| \leq \sum_{i=1}^{n} (t_n - t_{n-1}) = t_n \leq t^* ,$$

where the sequence $\{t_n\}$ is the same as that defined in Lemma 4.4, Chapter 2. Now, using that lemma we continue the same argument as in the proof of Theorem 3.1.

Let us discuss a special case of Theorem 5.1 where the two point contractor Γ is generated by the first divided difference functional G of F. Thus, we consider the iterative procedure

$$x_{n+1} = x_n - Fx_n \Gamma(x_n, \bar{x}_n) , \quad n = 0, 1, \cdots , \tag{5.6}$$

where $\Gamma(x_n, \bar{x}_n) = y_n / G(x_n, \bar{x}_n) y_n$ with y_n chosen so as to satisfy

$$G(x_n, \bar{x}_n) y_n \geq \beta \|G(x_n, \bar{x}_n)\| \text{ and } \|y_n\| = 1 , \tag{5.7}$$

where $\beta < 1$ is any positive fixed number. We assume also that G satisfies the following condition

$$\|G(x, \bar{x}) - G(\xi, \bar{\xi})\| \leq K[\|x - \xi\| + \|\bar{x} - \bar{\xi}\|] \tag{5.8}$$

for $x, \bar{x}, \xi, \bar{\xi} \in D$. Suppose that

$$1/\|G(x_0, \bar{x}_0)\| \leq \bar{B} \tag{5.9}$$

and

$$B = \bar{B}/\beta , \quad \bar{M} = K(1 + \gamma_2) , \quad B\bar{M}r \leq 1 .$$

Then we obtain

$$\begin{aligned} \|G(x, \bar{x})\| &\geq \|G(x_0, \bar{x}_0)\| \, |1 - [\|G(x_0, \bar{x}_0) - G(x, \bar{x})\|] / \|G(x_0, \bar{x}_0)\| \, | \\ &\geq \|G(x_0, \bar{x}_0)\| (1 - \bar{B}K[\|x - x_0\| + \|\bar{x} - \bar{x}_0\|]) \geq (1 - B\bar{M}\|x - x_0\|)/B \end{aligned}$$

by virtue of (5.8) and (5.9). Hence, we conclude from (5.7) that

$$\|\Gamma(x, \bar{x})\| \leq B/(1 - B\bar{M}\|x - x_0\|) \quad \text{for } x, \bar{x} \in S \text{ if } B\bar{M}r \leq 1 . \tag{5.10}$$

<u>Theorem 5.2</u>. Let $F : D \to R$ be a nonlinear functional with domain D

containing the ball $S = S(x_0,r)$. Suppose that the first divided difference functional G of F exists and satisfies conditions (5.8) and (5.9). Put $a = BK \max[2\gamma_1, 1+\gamma_2]$ and assume that conditions (5.3) and (5.4) are satisfied. Then all x_n lie in S and the sequence $\{x_n\}$ defined by (5.6), (3.4) and (3.5) converges to a root x of $Fx = 0$ and the error estimate (5.5) holds.

Proof. It follows from (5.8) that Γ is a two point contractor for F satisfying the contractor inequality (5.1). We infer from (5.10) that (5.2) is also satisfied with M replaced by BK . Hence, it follows that all hypotheses of Theorem 5.1 are satisfied and the proof follows immediately.

Consider now the iterative procedure defined by (5.6) and (1.7). We obtain the following theorem as a special case of Theorem 5.2.

Theorem 5.3. Under the hypotheses of Theorem 5.2, let the sequence $\{x_n\}$ be defined by (5.6), (1.7) and (3.5). If, in addition,

$$\|G(x,\bar{x})\| \leqq \delta \quad \text{for} \quad x,\bar{x} \in S , \tag{5.11}$$

then all assertions of Theorem 5.2 hold with $\gamma_1 = 1 + \delta C$.

Proof. It remains to prove that condition (3.4) is satisfied. This follows from Lemma 4.1 with $J(x,\bar{x})$ replaced by $G(x,\bar{x})$. Then we conclude that condition (3.4) is satisfied with $\gamma_1 = 1 + \delta C$.

CHAPTER 10

RELATIONS RESEARCH

Introduction. Many mathematical problems, in particular the problem of solving an operator equation can be described in terms of sets, points, and their relationships. For instance, the problem of existence of a solution to an operator equation is equivalent to the problem of whether a particular point (often 0) belongs to the range of the operator in question. One can say that we have here a point-set relationship. Thus, the problem of solvability of an operator equation is a point-set solvability problem in the sense described above. An operator can also be defined by its graph, that is, a subset of related elements of the product of the domain and the range of the operator. By using this terminology, the solvability problem can be described in a different way.

The concept of contractor directions can be extended to arbitrary sets in Banach spaces in a natural way. Using the method of contractor directions, some general facts can be established concerning sets, points and their relationships. An application to Banach modules and to the factorization problem in Banach algebras is given in Appendix 3.

1. Point-set solvability

Definition 1.1. Given a set Z of the Banach space X and a point $x \in Z$, we say that $\Gamma_x(Z)$ is a set of contractor directions for Z at x if there exists a positive number $q = q(Z) < 1$ which has the following property. For each $y \in \Gamma_x(Z)$, there exist a positive number $\epsilon = \epsilon(x,y) \leq 1$ and an element $\bar{x} \in Z$ such that

$$\|\bar{x}-x-\epsilon y\| \leq q\epsilon\|y\| . \tag{1.1}$$

This definition is a particular case of Definition 1.1, Chapter 5, where $P = I$ is the identity mapping of Z.

Lemma 1.1. The closure of $\Gamma_x(Z) = \Gamma_x(Z,q)$ is contained in $\Gamma_x(Z,\bar{q})$, where $0 < q < \bar{q} < 1$. If $\Gamma_x(Z)$ is dense in some ball $S(0,r) \subset X$, then $\Gamma_x(Z) = X$.

Proof. The proof follows from Lemma 1.1, Chapter 5, where $P = I$.

Lemma 1.2. Let Z be a closed set of the Banach space X and y_0 a point of X which is not in Z. Suppose that for each $x \in Z$, $y_0 - x$ is in the closure of the set $\Gamma_x(Z)$. Then there exists a sphere $S(y_0,r) \subset X$ such that, for all $y \in S(y_0,r)$ and $x \in Z$, $y-x$ belongs to a set $\Gamma_x(Z)$.

Proof. The proof follows from Lemma 1.2, Chapter 5, where $P = I$.

Theorem 1.1. Let Z be a closed set of the Banach space X, and $y_0 \in X$. Then $y_0 \in Z$ if and only if $y_0 - x$ belong to the closure of $\Gamma_x(Z)$ for all $x \in Z$. If y_0 is not on the boundary of Z and $y_0 - x$ belong to the closure of $\Gamma_x(Z)$ for all $x \in Z$, then there is a ball $S(y_0,r)$ with center y_0 and radius r which is contained in Z.

Proof. The proof follows from Theorem 1.1 or 6.1, Chapter 5, where $P = I$.

Theorem 1.2. Let Z be a closed set of the Banach space X. Then $Z = X$ if and only if the closure of $\Gamma_x(Z)$ is the whole of X for all x of X.

Proof. The proof follows from Theorem 1.2, Chapter 5, where P should be replaced by the identity mapping of Z.

Remark 1.1. Theorems 1.1 and 1.2 can be considered as examples of the point-set solvability problem. These theorems have been obtained in a formal way as special cases of more general theorems concerning the solvability of operator equations. On the other hand, the converse is also true. The solvability of the operator equation $Px = y_0$, $x \in X$,

$y_0 \in Y$, reduces to a point-set solvability problem by putting $Z = P(X)$, and the question is then whether y_0 belongs to the set Z . This is a reflection of the formal point of view. However, the point set solvability problem seems to be more general in a sense that situations may occur far beyond the traditional operator theory.

2. Compactness and contractor directions

Definition 2.1. An infinite set of the Banach space X is said to have the accretion property if there exists an element $y_0 \in X$ and a positive constant $q < 1$ which have the following property: for every $x \in Z$ there exists a positive number $\epsilon \leq 1$ and an element $\bar{x} \in Z$ such that $\bar{x} \neq y_0$ and

$$\|\bar{x}-x-\epsilon(y_0-x)\| \leq q\epsilon\|y_0-x\| . \tag{2.1}$$

Lemma 2.1. An infinite set Z of the Banach space X has the accretion property if and only if there exists a limit point of Z .

Proof. Let us prove that the accretion property is sufficient. Suppose that Z has no limit points in X . Then the set $Z_0 = Z\backslash\{y_0\}$ is closed and it follows from condition (2.1) that there exists a set $\Gamma_x(Z_0)$ of contractor directions for Z_0 at each $x \in Z_0$ such that

$$y_0-x \in \Gamma_x(Z_0) \quad \text{for all} \quad x \in Z_0 .$$

Hence, by virtue of Theorem 1.1, $y_0 \in Z_0$, which is a contradiction. The proof of necessity follows immediately. In fact, let $y_0 \in X$ be a limit point of Z and let q be an arbitrary positive number such that $q < 1$. Then, for each $x \in Z$, there exists an element $\bar{x} \in Z$ which satisfies condition (2.1) with $\epsilon = 1$ and $\bar{x} \neq y_0$.

Theorem 2.1. A set Z of the Banach space X is compact in X if and only if every infinite subset of Z has the accretion property.

Proof. The proof follows immediately from Lemma 2.1.

Another formulation of Theorem 2.1 can be given as follows.

Theorem 2.2. A set Z of the Banach space X is compact in X if and only if for every infinite sequence $\{x_n\}$ of elements of Z there exists a positive number $q < 1$ and an element $y \in X \setminus \{x_n\}$ which have the following property: A) for every x_i there exists a positive number $\epsilon_i \leq 1$ $(i = 1,2,\cdots)$ and an element x_j of $\{x_n\}$ such that

$$\|x_j - x_i - \epsilon_i (y - x_i)\| \leq q\epsilon_i \|y - x_i\| .$$

Let $P:X \to Y$ be a mapping of a Banach space X into a Banach space Y. As an application of Theorem 2.2, we obtain the following

Corollary 2.1. An operator $P:X \to Y$, where X and Y a Banach spaces is compact if and only if for every infinite bounded sequence $\{x_n\} \subset X$ there exists a positive number $q < 1$ and an element $y \in Y$ $y \in Y \setminus \{Px_n\}$ which have the following property. For every x_i $(i = 1,2,\cdots)$ there exists a positive number $\epsilon_i \leq 1$ and an element x_j of $\{x_n\}$ such that

$$\|Px_j - Px_i - \epsilon_i (y - Px_i)\| \leq q\epsilon \|y - Px_i\| .$$

Remark 2.1. What is actually proved in Theorem 2.2 is that y is a limit point of a sequence $\{x_n\}$ in a Banach space X if there exists a positive $q < 1$ which has the property A), and conversely.

3. R-solvability

Let X be an abstract set and let Y be a Banach space. Consider a subset Z of the product $X \times Y$, and denote the elements of Z by $\xi = (x,y)$, where $x \in X$, $y \in Y$. Suppose that there exists some relationship between the elements x and y which describes the subset $Z \subset X \times Y$. Then we say that $y \in Y$ has a related element in Z if there is an element $x \in X$ such that $(x,y) \in Z$. The set of all elements $y \in Y$ which have related elements in X is, obviously the projection of Z into Y and will be designated by $\mathrm{pr}_Y(Z)$.

The question under consideration is when does a particular element $y_0 \in Y$ have a related element in X. If we assume that the projection

$\underset{Y}{pr}(Z)$ is closed in Y, the necessary and sufficient conditions are given by the following theorem.

Theorem 3.1. Given a subset $Z \subset X \times Y$, where X is an abstract set and Y is a Banach space, suppose that the projection $\underset{Y}{pr}(Z)$ is closed in Y. Then $y_0 \in Y$ has a related element $x_0 \in X$ if and only if there exists a positive constant $q < 1$ which has the following property. For each pair $(x,y) \in Z$, there exists a positive number $\epsilon \leq 1$ and a pair $(\bar{x},\bar{y})$ such that

$$\|\bar{y}-y-\epsilon(y_0-y)\| \leq q\epsilon\|y_0-y\| . \tag{3.1}$$

Proof. The proof is the same as that of Theorem 1.1, Chapter 5. Moreover, let us observe that Theorem 1.1, Chapter 5, also holds true for a multi-valued mapping $P:X \to Y$. Therefore, the proof of Theorem 3.1 can be reduced to the multi-valued case.

Now let X be a complete metric space and let Y be a Banach space. Consider the question of existence of related elements without assumption that the projection $\underset{Y}{pr}(Z)$ of $Z \subset X \times Y$ is closed in Y. The following theorem gives only sufficient conditions.

Theorem 3.2. Let Z be a closed subset of $X \times Y$ (in the natural product topology). Then $y_0 \in Y$ has a related element $x_0 \in X$ if there exists a positive constant $q < 1$ and a continuous increasing function $B(s) > 0$ for $s > 0$ which satisfies condition (2.1), Chapter 5, i.e., $B \in \mathbb{B}$, with the following property: for each pair $(x,y) \in Z$ there exists a positive number $\epsilon \leq 1$ and a pair $(\bar{x},\bar{y})$ such that condition (3.1) is satisfied and

$$d(\bar{x},x) \leq \epsilon B(\|y_0-y\|) . \tag{3.2}$$

Proof. The proof is the same as that of Theorem 2.1, Chapter 5. On the other hand, it is easily seen that Theorem 2.1, Chapter 5, also applies to a multi-valued mapping $P:X \to Y$. Therefore, the proof of Theorem 3.2 can be reduced to the multi-valued case.

Let $P:X \to Y$ be in general a multi-valued mapping of a complete metric space X into a Banach space Y. The solvability of the operator equation $Px = y$ is a classical example of an R-solvability problem. The graph of the mapping P determines the subset Z of $X \times Y$. The operator P defines a relationship between the elements x and $y = Px$ of a pair (x,y) of Z. Suppose that the graph of P is closed, then we have a situation described by Theorem 3.2. The case where the range of P is closed is described by Theorem 3.1. In this case, the metric space X can be replaced by an abstract set. Theorems 1.1 and 2.1, Chapter 5, are model theorems for Theorems 3.1 and 3.2, respectively. Inequalities (2.2) and (2.3), Chapter 5, are counterparts of inequalities (3.1) and (3.2), respectively.

The implicit function theorem provides another example of an R-solvability problem. In fact, let $F:X \times Y \to E$ be a mapping, then the subset Z of $X \times Y$ is defined by the relationship $F(x,y) = 0$, where (x,y) are the elements of Z. Given a subset U of X, then $y = \varphi(x)$ is related to $x \in U$ if $F(x,\varphi(x)) = 0$.

4. Intersection of sets

4.1. Let Z be a closed subset of the product space $X \times X$, where X is a Banach space. By using the technique of contractor directions, sufficient conditions can be given which guarantee the existence of at least one element of Z being of the form (x,x). Such an element will be called a diagonal element of Z.

Theorem 4.1. Suppose that there exists a positive constants $q < 1$ and a function $B \in \mathbb{B}$ which have the following property. For each pair (x,y) of Z, there exists a pair $(\bar{x},\bar{y})$ and a positive $\epsilon \leq 1$ such that

$$\|(\bar{x}-\bar{y}) - (1-\epsilon)(x-y)\| \leq q\epsilon\|x-y\| \tag{4.1}$$

and

$$\|\bar{x}-x\| \leq \epsilon B(\|x-y\|) . \tag{4.2}$$

Then Z has at least one diagonal element $(x,x) \in Z$.

Proof. The proof is similar to that of Theorem 2.1 of Chapter 5. Using Greek letters for ordinal numbers of first and second number class, we construct well-ordered sequences of numbers t_α and pairs $(x_\alpha, y_\alpha) \in Z$ as follows, where for simplicity we put $B(s) = Bs$ for some $B > 0$.

Put $t_0 = 0$ and let (x_0, y_0) be an arbitrary pair from Z . Suppose that t_γ and $(x_\gamma, y_\gamma) \in Z$ have been constructed for all $\gamma < \alpha$, provided: for arbitrary ordinal numbers $\gamma < \alpha$, inequality (4.3_γ) is satisfied.

$$\|z_\gamma\| \le e^{-(1-q)t_\gamma}\|z_0\| , \tag{4.3$_\gamma$}$$

where $z_\gamma = x_\gamma - y_\gamma$. For first kind ordinal numbers $\beta = \gamma+1 < \alpha$, the following inequalities are satisfied:

$$0 < t_{\gamma+1} - t_\gamma \le 1 , \tag{4.4$_{\gamma+1}$}$$

$$\|x_{\gamma+1} - x_\gamma\| \le B\|z_0\|e^{-(1-q)t_\gamma}(t_{\gamma+1} - t_\gamma) , \tag{4.5$_{\gamma+1}$}$$

$$\|z_{\gamma+1} - z_\gamma\| \le (1+q)\|z_0\|e^{-(1-q)t_\gamma}(t_{\gamma+1} - t_\gamma) ; \tag{4.6$_{\gamma+1}$}$$

and for second kind (limit) ordinal numbers $\gamma < \alpha$ the following relations hold:

$$t_\gamma = \lim_{\beta \nearrow \gamma} t_\beta , \quad x_\gamma = \lim_{\beta \nearrow \gamma} t_\beta , \quad z_\gamma = \lim_{\beta \nearrow \gamma} z_\beta , \tag{4.7$_\gamma$}$$

where $z_\beta = x_\beta - y_\beta$. Then it follows from Lemmas 3.1 and 3.2, Chapter 4 that for arbitrary $\lambda < \gamma < \alpha$ we have, by virtue of (4.6), (4.4) and (4.7), that

$$\|z_\gamma - z_\lambda\| \le \sum_{\lambda \le \beta < \gamma} \|z_{\beta+1} - z_\beta\| \le (1+q)\|z_0\| \sum_{\lambda \le \beta < \gamma} e^{-(1-q)t_\beta}(t_{\beta+1} - t_\beta) =$$

$$= (1+q)\|z_0\| \sum_{\lambda \le \beta < \gamma} e^{(1-q)(t_{\beta+1} - t_\beta)} e^{-(1-q)t_{\beta+1}}(t_{\beta+1} - t_\beta) <$$

$$< (1+q)\|z_0\|e^{1-q} \sum_{\lambda \le \beta < \gamma} e^{-(1-q)t_{\beta+1}}(t_{\beta+1} - t_\beta) <$$

$$< (1+q)e^{1-q}\|z_0\| \sum_{\lambda\le\beta<\gamma} \int_{t_\beta}^{t_{\beta+1}} e^{-(1-q)t}dt = (1+q)e^{1-q}\|z_0\| \int_{t_\lambda}^{t_\gamma} e^{-(1-q)t}dt .$$

Hence,

$$\|z_\gamma - z_\lambda\| \le (1+q)e^{1-q}\|z_0\| \int_{t_\lambda}^{t_\gamma} e^{-(1-q)t} dt . \tag{4.8}$$

In the same way we obtain from (4.4), (4.5) and (4.7) that

$$\|x_\gamma - x_\lambda\| \le Be^{1-q}\|z_0\| \int_{t_\lambda}^{t_\gamma} e^{-(1-q)t}dt . \tag{4.9}$$

Suppose that α is a first kind ordinal number. If $z_{\alpha-1} = x_{\alpha-1} - y_{\alpha-1} = 0$, then the proof is completed. If $z_{\alpha-1} \ne 0$, then for $(x_{\alpha-1}, y_{\alpha-1}) \in Z$, there exists a pair $(\bar{x},\bar{y}) \in Z$ and a positive $\epsilon \le 1$ which satisfy (4.1) and (4.2) with $x = x_{\alpha-1}$ and $y = y_{\alpha-1}$, and we put

$$t_\alpha = t_{\alpha-1} + \tau_\alpha , \quad x_\alpha = \bar{x} , \quad y_\alpha = \bar{y} , \tag{4.10}$$

where $\epsilon = \tau_\alpha$. Thus, we have

$$\|z_\alpha - (1-\tau_\alpha) z_{\alpha-1}\| \le q\tau_\alpha \|z_{\alpha-1}\| . \tag{4.11}$$

Hence, we obtain, by (4.11) and ($4.3_{\alpha-1}$),

$$\|z_\alpha\| \le (1-\tau_\alpha)\|z_{\alpha-1}\| + \tau_\alpha\|z_{\alpha-1}\| = (1-(1-q)\tau_\alpha)\|z_{\alpha-1}\| <$$
$$< e^{-(1-q)\tau_\alpha}\|z_{\alpha-1}\| \le e^{-(1-q)t_\alpha}\|z_0\| ,$$

i.e.,

$$\|z_\alpha\| \le e^{-(1-q)t_\alpha}\|z_0\| . \tag{4.12}$$

This means that the induction assumption (4.3_γ) is also satisfied for $\gamma = \alpha$. It follows from (4.10), (4.1), (4.2) and ($4.3_{\alpha-1}$) that

$$\|x_\alpha - x_{\alpha-1}\| \le B\tau_\alpha\|z_{\alpha-1}\| \le B(t_\alpha - t_{\alpha-1})e^{-(1-q)t_{\alpha-1}}\|z_0\| . \tag{4.13_α}$$

Similarly, it follows from (4.10), (4.11) and ($4.3_{\alpha-1}$) that

$$\|z_\alpha - z_{\alpha-1}\| \le (1+q)\tau_\alpha\|z_{\alpha-1}\| \le (1+q)\|z_0\|e^{-(1-q)t_{\alpha-1}}(t_{\alpha-1}) . \tag{4.14_α}$$

Thus, conditions (4.3_α) - (4.6_α) are satisfied for t_α , x_α and

$z_\alpha = x_\alpha - y_\alpha$. Now suppose that α is an ordinal number of second kind and put $t_\alpha = \lim_{\gamma \nearrow \alpha} t_\gamma$. Let $\{\gamma_n\}$ be an increasing sequence convergent to α . It follows from (4.8) and (4.9) that $\{z_{\gamma_n}\}$ and $\{x_{\gamma_n}\}$ are Cauchy sequences, and so are $\{x_\gamma\}$ and $\{z_\gamma\}$. Denote by x_α and z_α their limits, respectively. Since $z_\gamma = x_\gamma - y_\gamma$, the sequence $\{y_\gamma\}$ has a limit $y_\alpha = x_\alpha - z_\alpha$. Since Z is closed, we infer that $(x_\alpha, y_\alpha) \in Z$. If $t_\alpha < +\infty$, then the limit passage in (4.3_{γ_n}) yields (2.3_α). The relationships (4.7_α) are satisfied by the definition of t_α , x_α , y_α and z_α . This process will terminate if $t_\alpha = +\infty$, where α is an ordinal number of second kind. In this case, $z_\alpha = 0$, by virtue of (4.3_α). Finally, we conclude that $x_\alpha = y_\alpha$ and $(x_\alpha, x_\alpha) \in Z$. Thus, the proof is completed.

4.2. In order to investigate the problem of existence of diagonal elements of a closed subset of a product space of a Banach space by itself, an argument has been applied which is actually based on the concept of contractor directions. This leads in a natural way to the idea of exploiting the same argument in order to investigate the intersection of two closed subsets of a Banach space.

Given two closed subsets X and Y of the Banach space E , the problem is to explore the case when their intersection is nonempty, that is $X \cap Y \neq \emptyset$ (empty set).

Theorem 4.2. Let X and Y be two closed subsets of the Banach space E . Suppose that there exist positive constants B and $q < 1$ with the following property. For each pair of elements $x \in X$ and $y \in Y$, there exists a pair of elements $\bar{x} \in X$ and $\bar{y} \in Y$ and a positive number $\epsilon \leq 1$ such that

$$\|(\bar{x}-\bar{y}) - (1-\epsilon)(x-y)\| \leq q\epsilon\|x-y\| \tag{4.15}$$

and

$$\|\bar{x}-x\| \leq B\epsilon\|x-y\| . \tag{4.2}$$

Then $X \cap Y \neq \emptyset$.

Proof. We construct well-ordered sequences of numbers t_α and elements $x_\alpha \in X$, $y_\alpha \in Y$ and $z_\alpha = x_\alpha - y_\alpha$ as in the proof of Theorem 4.1 and we show that $z_\alpha \to 0$ and $x_\alpha \to x \in X$ and $y_\alpha \to y \in Y$.

Assume that one of the subsets in question is compact. Then a necessary and sufficient condition for nonemptiness of the intersection is given by the following theorem.

Theorem 4.3. Let X and Y be two closed subsets of the Banach space E. Suppose, in addition, that X is compact. Then $X \cap Y \neq \emptyset$ if and only if there exists a positive constant $q < 1$ with the following property. For each pair of elements $x \in X$ and $y \in Y$, there exists a pair of elements $\bar{x} \in X$ and $\bar{y} \in Y$ and a positive number $\epsilon \leq 1$ such that (4.15) is satisfied.

Proof. The proof that condition (4.15) is necessary is immediate. In fact, let x^* be in $X \cap Y$, and let $0 < q < 1$ be arbitrary but fixed. Then condition (4.15) will be satisfied for an arbitrary pair of elements $x \in X$ and $y \in Y$ if we put $\epsilon = 1$, and $\bar{x} = \bar{y} = x^*$. The proof that condition (4.15) is sufficient is almost the same as that of Theorem 4.2 which in turn imitates the proof of Theorem 4.1. The only change to be made is in the induction assumption (4.7_γ), where $x_\gamma = \lim_{\beta \nearrow \gamma} x_\beta$ should be replaced by "x_γ is a limit point of the sequence $\{x_\beta\}$". This is because the induction assumption $(4.5_{\gamma+1})$ has to be dropped, since we do not assume (4.2). Then y_γ will be the limit of the corresponding subsequence of $y_\beta = x_\beta + z_\beta$. Finally, if $t_\alpha = +\infty$, then $\lim_{\beta \nearrow \alpha} z_\beta = 0$. Suppose that x_α is a limit point of $\{x_\beta\}$ and y_α is the limit of the corresponding subsequence of $\{y_\beta\}$, then $y_\alpha = x_\alpha$.

4.3. Let Z be a closed subset of the n-product space $X \times X \times \ldots \times X$, where X is a Banach space and denote by $(x^1, x^2, \cdots, x^n)$ the elements of Z. The problem of existence of diagonal elements $(x, x, \cdots, x)$ of Z, i.e., $x^i = x$ for $i = 1, 2, \cdots, n$, can also be investigated by using the same technique.

Theorem 4.4. Suppose that there exist positive constants B and $q < 1$ with the following property. For each n-tuple $(x^1, x^2, \cdots, x^n)$ of Z there exists a n-tuple $(\bar{x}^1, \bar{x}^2, \cdots, \bar{x}^n) \in Z$ and a positive $\epsilon \leq 1$ and an index k with $1 \leq k \leq n$ such that

$$(4.16) \qquad \|(\bar{x}^i - \bar{x}^k) - (1-\epsilon)(x^i - x^k)\| \leq q\epsilon \|x^i - x^k\|$$

for $i \neq k$ and

$$(4.17) \qquad \|\bar{x}^i - x^i\| \leq B\epsilon \|x^i - x^k\| \quad \text{for} \quad i \neq k .$$

Then Z has at least one diagonal element $(x, x, \cdots, x) \in Z$.

Proof. The proof follows the pattern of the proof of Theorem 4.1. Put $x = (x^1, x^2, \cdots, x^n) \in Z$ with norm $\|x\| = \sum_{i=1}^{n} \|x^i\|$ and $z = (x^1 - x^k, x^2 - x^k, \cdots, x^n - x^k)$. Similarly, put $\bar{z} = (\bar{x}^1 - \bar{x}^k, \bar{x}^2 - \bar{x}^k, \ldots, \bar{x}^n - \bar{x}^k)$, $\bar{x} = (\bar{x}^1, \bar{x}^2, \cdots, \bar{x}^n) \in Z$, where k depends on x and is defined by (4.16), (4.17). Then inequalities (4.16), (4.17) can be written in the form

$$(4.18) \qquad \|\bar{z} - (1-\epsilon)z\| \leq q\epsilon \|z\|$$

$$(4.19) \qquad \|\bar{x} - x\| \leq B\epsilon \|z\| .$$

We construct well-ordered sequences of numbers t_α and n-tuples x_α , z_α . Now the proof proceeds formally as that of Theorem 4.1. Thus, in the induction assumptions (4.3) - (4.7), the elements x_α and z_α are to be replaced by the corresponding n-tuples x_α and z_α defined above. In relationships (4.10), (4.11), we put $x_\alpha = \bar{x}$ and $z_\alpha = \bar{z}$, $\tau_\alpha = \epsilon$, where $\bar{x}$, $\bar{z}$ and ϵ are determined by (4.16) and (4.17) or, equivalently, by (4.18) and (4.19). In the conclusion of the proof, we obtain sequences $t_\gamma \to t_\alpha = +\infty$, $x_\gamma \to x_\alpha$ and $z_\gamma \to z_\alpha = 0$, where α is an ordinal number of second kind. Since k in (4.16), (4.17) assumes a finite number of values $(1 \leq k \leq n)$, there exists an infinite subsequence $\{z_{\gamma_m}\}$ of $\{z_\gamma\}$ and a fixed k such that

$z_{\gamma_m} = (x^1_{\gamma_m} - x^k_{\gamma_m}, x^2_{\gamma_m} - x^k_{\gamma_m}, \cdots, x^n_{\gamma_m} - x^k_{\gamma_m})$. Since $x_\alpha = (x^1_\alpha, x^2_\alpha, \cdots, x^n_\alpha)$ is the limit of the sequence $\{x_{\gamma_m}\}$ and the limit of $\{z_{\gamma_m}\}$ is $z_\alpha = (0,0,\cdots,0)$, it follows that $x^i_\alpha = x^k_\alpha$ for $i = 1,2,\cdots,n$, and, consequently, $(x^k_\alpha, x^k_\alpha, \cdots, x^k_\alpha) \in Z$. This completes the proof of the theorem.

4.4. Given n closed subsets, X^i $(i = 1,2,\cdots,n)$ of the Banach space E , the problem is now to investigate the case when $\bigcap_{i=1} X^i \neq \emptyset$. The following theorem gives only sufficient conditions.

Theorem 4.5. Given n closed subsets X^i $(i = 1,2,\cdots,n)$ of the Banach space E , suppose that there exist positive constants B and $\epsilon \leq 1$ with the following property. For arbitrary n elements $x^i \in X$ $(i = 1,2,\cdots,n)$, there exist n elements $\bar{x}^i \in X^i$ $(i = 1,2,\cdots,n)$ and a positive number $\epsilon \leq 1$ and an index k with $1 \leq k \leq n$ such that

$$(4.20) \qquad \|(\bar{x}^i - \bar{x}^k) - (1-\epsilon)(x^i - x^k)\| \leq q\epsilon \|x^i - x^k\| \quad \text{for } i \neq k ,$$

and

$$(4.21) \qquad \|\bar{x}^i - x^i\| \leq B\epsilon \|x^i - x^k\| \quad \text{for } i \neq k .$$

Then $\bigcap_{i=1}^{n} X^i \neq \emptyset$.

Proof. As in the proof of Theorem 4.4, we construct well-ordered sequences of numbers t_α and elements $x^i_\alpha \in X^i$, $z^i_\alpha = x^i_\alpha - x^k_\alpha$ $(i = 1,2,\cdots,n)$, where k depends on $x^1_\alpha, x^2, \cdots, x^n_\alpha$ and its choice is made according to (4.20) and (4.21). The proof now follows the pattern of that of Theorem 4.4.

If the closed subsets in question are compact, then necessary and sufficient conditions for the nonemptiness of the intersection are given by the following theorem.

Theorem 4.6. Let X^i , $i = 1,2,\cdots,n$ be n closed compact subsets of the Banach space E . Then $\bigcap_{i=1}^{n} X^i \neq \emptyset$ if and only if there exists a positive constant $q < 1$ with the following property. For arbitrary n elements $x^i \in X$ $(i = 1,2,\cdots,n)$, there exist n elements $\bar{x}^i \in X^i$ $(i = 1,2,\cdots,n)$ and a positive number $\epsilon \leq 1$ and an index k with $1 \leq k \leq n$ such that condition (4.20) is satisfied.

Proof. The proof that condition (4.20) is necessary is immediate. In fact, let $x^* \in \bigcap_{i=1}^{n} X^i$ and let q be arbitrary with $0 < q < 1$. Then for arbitrary $x^i \in X^i$ $(i = 1,2,\cdots,n)$ condition (4.20) is satisfied if we put $\bar{x}^i = x^*$ for $i = 1,2,\cdots,n$ and $\epsilon = 1$. The proof that condition (4.20) is sufficient is almost the same as that of Theorem 4.5 which in turn imitates the proof of Theorem 4.4. Instead of condition (4.21), we use the compactness of the subsets X^i $(i = 1,2,\cdots,n)$ in order to replace the limits $x_\gamma = \lim_{\beta \nearrow \gamma} x_\beta$ which may not exist by limit points of the sequences $\{x_\beta\}$ (see the corresponding remark in the proof of Theorem 4.3).

Remark 4.1. Condition (4.17) in Theorem 4.4 and condition (4.21) in Theorem 4.5 can be replaced by the following one

$$\|\bar{x}^i - x^i\| \leq \epsilon B(\|x^i - x^k\|) \quad \text{for} \quad i \neq k ,$$

where $B(s) > s$ is a continuous increasing function such that $\int_0^1 s^{-1} B(s)ds < \infty$ (see Chapter 3).

APPENDIX 1

A STRATEGY THEORY OF SOLVING EQUATIONS

Introduction. The strategy theory to be discussed is actually a part of the general theory of contractor directions presented in Chapters 5 and 6. We have shown there the structure of the general theory which is such that it takes all the benefits of the abstract differential calculus whereas no differentiation of any kind is involved in the general definition of contractor directions. Moreover, there are cases of operators which are not differentiable and which nevertheless can be handled by the method of contractor directions. This is, for instance, the case of contraction mappings. There is also another advantage of the general theory, namely, if the range of the operator involved in the equation is closed, then the method of contractor directions provides solvability conditions which are both necessary and sufficient. As opposed to the general case, the strategy theory is based upon a sort of directional differentiability of rather special kind.

1. Strategic directions

Definition 1.1. Let $P:X \to Y$ be a nonlinear mapping of the vector space X into the Banach space Y. If $h \in X$ is such that

$$(1.1) \qquad \|P(x+\epsilon h) - (1-\epsilon)Px\|/\epsilon \to 0 \quad \text{as} \quad \epsilon \to 0+ ,$$

then h is called a strategic direction for P at $x \in X$.

It follows from this definition that $-Px \in \Gamma_x(P,q)$ for arbitrary positive $q < 1$, i.e., $-Px$ is a contractor direction for P at x.

Definition 1.2. If, for every $x \in X$, there exists an element $\sigma(x) \in X$ such that

(1.2) $$\|P(x+\epsilon\sigma(x)) - (1-\epsilon)Px\|/\epsilon \to 0 \quad \text{as} \quad \epsilon \to 0+ ,$$

then $\sigma:X \to X$ is called a strategic mapping for P or briefly a strategy for P .

Suppose that P is differentiable in the Fréchet sense, then it follows from (1.2) that

$$P'(x)\sigma(x) = -Px \quad \text{for all} \quad x \in X .$$

If, in addition, $P'(x)$ is not singular, then we have

(1.3) $$\sigma(x) = -P'(x)^{-1}Px , \quad x \in X .$$

Hence, it follows that if P possesses a nonsingular Fréchet derivative, then P has a strategy which is defined by the formula (1.3), and, obviously this is the strategy involved in the well-known Newton-Kantorovič method (see (2.11), Chapter 1).

<u>Theorem 1.1</u>. Let $P:X \to Y$ be a nonlinear mapping of the vector space X into the Banach space Y . Suppose that the range $P(X)$ is closed. If P has a strategy, then the equation

(1.4) $$Px = 0 , \quad x \in X ,$$

has a solution.

<u>Proof</u>. The proof follows from Theorem 1.1, Chapter 5, since $-Px \in \Gamma_x(P,q)$ for arbitrary positive $q < 1$ and all $x \in X$.

As a particular case of Theorem 1.1, we obtain the following

<u>Theorem 1.2</u>. Let $P:X \to Y$ be a nonlinear mapping of the vector space X into the Banach space Y . Suppose that the Gâteaux derivative P' exists and is such that the linear equation

(1.5) $$P'(x)h = -Px , \quad x,h \in X$$

has a solution h for every $x \in X$. If the range $P(X)$ is closed, then equation (1.4) has a solution.

<u>Proof</u>. The proof follows immediately from Theorem 1.1.

2. Classified strategic directions

We now discuss the problem of solving equation (1.4) in the general case, i.e., without assumption that the range $P(X)$ is closed.

<u>Definition 2.1</u>. Let $P:D(P) \subset X \to Y$ be a nonlinear mapping, where $D(P)$ is a vector space and X, Y are Banach spaces. Let $B(s) > 0$ for $s > 0$ be continuous increasing function such that $\int_0^1 s^{-1}B(s)ds < \infty$, and let $g(t) > 0$ for $t > 0$ be a continuous function. If there exists an element $h \in D(P)$ which satisfies condition (1.1) and

$$\|h\| \leq B(g(\|x\|)\cdot\|Px\|) ,$$

then h is a strategic direction with the (B,g)-property for the operator P at $x \in D(P)$.

Hence, it follows that Definition 2.1 is a special case of Definition 1.1.

<u>Definition 2.2</u>. If, for every $x \in D(P)$, there exists an element $\sigma(x) \in D(P)$ which satisfies condition (1.2) and

$$\|\sigma(x)\| \leq B(g(\|x\|)\cdot\|Px\|) , \tag{2.1}$$

then $\sigma:D(P) \to D(P)$ is called a strategic mapping with the (B,g)-property or briefly a strategy with the (B,g)-property, or a (B,g)-strategy.

Hence, it follows that Definition (2.2) is a special case of Definition 1.2. If $g(t) \equiv 1$, then we use the notation $(B,1)$.

<u>Theorem 2.1</u>. Let $P:D(P) \subset X \to Y$ be a closed mapping, where $D(P)$ is a vector space and X, Y are Banach spaces. If P has a $(B,1)$-strategy, then equation (1.4) has a solution.

<u>Proof</u>. It follows from Definition 2.2 that for all $x \in D(P)$, $-Px \in \Gamma_x(P)$, the set of contractor directions with the (B,I)-property. Thus, the hypotheses of Theorem 2.1, Chapter 5, are satisfied and the

proof follows immediately.

As a particular case of Theorem 2.1, we obtain the following

Theorem 2.2. Let $P:D(P) \subset X \to Y$ be a closed nonlinear mapping, where $D(P)$ is a vector space and X, Y are Banach spaces. Suppose that the Gâteaux derivative $P'(x)$ exists, and for every $x \in D(P)$, the equation (1.5) has a solution $\sigma(x)$ which satisfies the following condition

$$\|\sigma(x)\| \leq B(\|Px\|) , \tag{2.2}$$

where B is some function with the properties mentioned in Definition 2.1. Then equation (1.4) has a solution.

Proof. It results from (2.2) that condition (2.1) is satisfied with $g(t) \equiv 1$. Hence, it follows that P has a $(B,1)$-strategy, and the hypotheses of Theorem 2.1 are fulfilled.

Remark 2.1. In Theorem 1.1, Chapter 6, condition (1.4) can be replaced by the assumption that P has a (B,q)-strategy.

Similar remarks are also valid for other theorems concerning the existence of a solution of equation (1.4).

3. A necessary and sufficient condition for the existence of a fixed point

Let Z be a closed subset of the Banach space X and let $F:Z \to Z$ be a continuous mapping such that $F(Z)$ is compact in X. The problem is to find necessary and sufficient conditions which guarantee the existence of a fixed point for F. Let us note that we do not assume that Z is convex. The following theorem solves the problem.

Theorem 3.1. Let $F:Z \to Z$ be a continuous mapping of the closed subset Z of the Banach space X such that the set $F(Z)$ is compact in X. Then F has a fixed point if and only if there exists a positive constant $q < 1$ with the following property. For each x of Z, there exists an element $\bar{x}$ of Z and a positive number $\epsilon \leq 1$ such that

$$\|\bar{x} - F\bar{x} - (1-\epsilon)(x-Fx)\| \leq q\epsilon \|x-Fx\| . \tag{3.1}$$

Proof. The proof that condition (3.1) is necessary follows immediately. In fact, suppose that x^* is a fixed point for F, i.e., $x^* = Fx^*$. Then for arbitrary $x \in Z$, condition (3.1) is satisfied if we put $\bar{x} = x^*$ and $\epsilon = 1$, where q is an arbitrary fixed number with $0 < q < 1$. The proof that condition (3.1) is sufficient follows from Theorem 1.1, Chapter 5. In fact, let $P:Z \to X$ be a mapping defined by $Px = x-Fx$. Since $F(Z)$ is compact, it follows that the range $P(Z)$ is closed in X. Condition (3.1) means precisely that $y_0-Px \in \Gamma_x(P,q)$ for all x of Z, where $y_0 = 0$ and $\Gamma_x(P,q)$ is a set of contractor directions for P at x. Thus, the hypotheses of Theorem 1.1, Chapter 5, are satisfied and, consequently, the equation $Px = 0$ has a solution x^*, that is, $x^*-Fx^* = 0$ and $x^* \in Z$. This completes the proof.

4. An implicit function theorem by means of contractor directions

Local existence theorems in terms of contractor directions can be used, in general, in order to prove implicit function theorems. To illustrate this fact, we use Theorem 2.1, Chapter 5, as a basis. Let X, Z and Y be Banach spaces and put

$$S = [(x,z) : \|x-x_0\| \leq r, \|z-z_0\| \leq \rho, x \in X, z \in Z]$$

for given $x_0 \in X$, $z_0 \in Z$, r and ρ.

Let $P:S \to Y$ be a continuous nonlinear operator and suppose that there exist sets $\Gamma_\xi(P)$ of(z-uniform)contractor directions in the following sense where $\xi = (x,z) \in S$. There exists a positive $q < 1$ which has the following property. For arbitrary $y \in Y$ and $\xi = (x,z) \in S$, there exist $h(z) \in X$ and a positive $\epsilon \leq 1$ such that (ϵ is independent of z and)

$$\|P(x+\epsilon h(z),z) - P(x,z) - \epsilon y\| \leq q\epsilon \|y\| , \tag{4.1}$$

where ($h(z)$ is continuous in z, and)

$$(4.2) \qquad \|h(z)\| \le B(\|y\|) ,$$

where $B(s) > 0$ for $s > 0$ is some continuous increasing function such that $\int_0^1 s^{-1}B(s)ds < \infty$. Let us notice that in (4.1) and (4.2), it is sufficient to assume $y = -P(x,z)$.

Theorem 4.1. Suppose that $P:S \to Y$ is a continuous mapping satisfying the following conditions:

1) $P(x_0,z_0) = 0$

2) P is endowed with sets $\Gamma_\xi(P)$ of z-uniform contractor directions satisfying conditions (4.1) and (4.2). Then there exists a continuous function $g(z)$ defined in some neighborhood of z_0, with values in X, such that $P(g(z),z) = 0$.

Proof. First of all we choose η and ρ_1 such that

$$\|P(x_0,z)\| \le \eta \quad \text{for} \quad \|z-z_0\| \le \rho_1$$

and

$$(1-q)^{-1}\int_0^a s^{-1}B(s)ds \le r \quad \text{with} \quad a = e^{1-q}\eta .$$

Now in the same way as in the proof of Theorem 2.1, Chapter 5, we construct sequences of positive numbers t_α ($t_0=0$) and continuous functions $x_\alpha(z)$ (replacing x_α in the mentioned theorem), where $(x_\alpha,z) \in S_1$

$$S_1 = [(x,z) : \|x-x_0\| \le r, \ \|z-z_0\| \le \rho_1] \subset X \times Z .$$

The values of $x_\alpha(z)$ lie in X and $x_0(z) \equiv x_0$. These sequences are to satisfy induction assumptions (2.4_γ), $(2.5_{\gamma+1})$ - (2.8_γ) provided that x_γ, $x_{\gamma+1}$, Px_γ, $Px_{\gamma+1}$ and Px_0 in Chapter 5 are replaced by $x_\gamma(z)$, $x_{\gamma+1}(z)$, $P(x_\gamma(z),z)$, $P(x_{\gamma+1}(z),z)$ and η, respectively. Thus, we obtain, by virtue of (4.1) and (4.2), the following estimates

$$(4.3) \qquad \|x_\gamma(z) - x_\lambda(z)\| \le \int_{t_\lambda}^{t_\gamma} B(e^{1-q}\, e^{-(1-q)t})dt$$

and

$$(4.4) \qquad \|P(x_\gamma(z),z) - P(x_\lambda(z),z)\| \leq (1+q)e^{1-q_\eta} \int_{t_\lambda}^{t_\gamma} e^{-(1-q)t}dt ,$$

replacing (2.9) and (2.10), Chapter 5. Inequalities (4.3) and (4.4) show the z-uniform convergence of the sequences of functions yielding the continuity of the limit functions. It follows from (4.3) that

$$\|x_\gamma(z) - x_0\| \leq (1-q)^{-1} \int_0^a s^{-1}B(s)ds \leq r ,$$

where $a = e^{1-q_\eta}$. The last inequality shows that $(x_\gamma(z),z) \in S_1$, where $\|z-z_0\| \leq \rho_1$. Thus, all constructed functions $x_\gamma(z)$ are continuous and well-defined. The further reasoning is exactly the same as in the proof of Theorem 2.1, Chapter 5.

Since the assumptions of Theorem 4.1 are rather weak, we cannot prove the uniqueness of the function $g(z)$.

APPENDIX 2

CONTRACTORS AND EQUATIONS IN PSEUDOMETRIC SPACES

Introduction. The concept of contractors has been studied in Chapters 1 - 3 as a tool for solving general equations in Banach spaces. By using the contractor method various existence theorems for solutions of equations are there obtained as well as convergence theorems for a broad class of iterative procedures. It is also shown that the contractor method yields a unified approach to a large variety of iterative processes entirely different in nature. A further development of the contractor concept is presented here. It is our aim to give an extension of this method to linear pseudometric spaces. For this purpose two schemes are suggested for the contractor method. As far as the majorant operator is concerned a special case of the first scheme is described in the monograph by Krasnosel'skii and others [2]. Regarding the second scheme, in the monograph by Collatz [1], a special case is presented which goes back to Schröder [1]. In the theory of iterative methods, by using partial orderings important results have been obtained by Kantorovič (for references, see Kantorovič and Akilov [1], see also Kurpel' [1]).

1.2. Following Collatz [1], a space X is called pseudometric if there is a mapping $\rho: X \times X \to E$, where E is a partially ordered linear space with notion of linear convergence. The pseudodistance $\rho(f,g)$ is assumed to have the following properties:

a) $\rho(f,g) \geq \theta$ (null element in E) .

b) $\rho(f,g) = \theta$ if and only if $f = g$.

c) $\rho(f,g) \leq \rho(f,h) + \rho(g,h)$ for arbitrary $f,g,h \in X$.

A sequence $\{f_n\} \subset X$ is called convergent to $f \in X$ that is $f_n \to f$ if $\rho(f_n,f) \to \theta$ as $n \to \infty$. Since we consider here only linear pseudometric spaces X , we assume that

$$\rho(x,x') = \rho(x-x',\theta_X) \quad \text{for all} \quad x,x' \in X \; ,$$

hence, $f_n \to f$ means $\rho(f_n-f,\theta_X) \to \theta$ as $n \to \infty$. The null element θ will be used without subscript. A sequence $\{f_n\} \subset X$ is said to be a Cauchy sequence if

$$\rho(f_n,f_m) \to \theta \quad \text{as} \quad m,n \to \infty \; .$$

We assume that all linear pseudometric spaces considered here are sequentially complete, that is, each Cauchy sequence in X has a limit element in X. The partially ordered space E_X associated with X is briefly called the metrizing space. For more details concerning pseudometric spaces, see Collatz [1].

2. Contractors with operator type majorants

Let X and Y be two linear pseudometric spaces with metrizing spaces E_X and E_Y, respectively. For arbitrary $x',x'' \in X$ let $\rho(x',x'') \in E_X$ and $\sigma(y',y'') \in E_Y$ for arbitrary $y',y'' \in Y$, ρ and σ stand for the metrics.

Let $P:X \to Y$ be a nonlinear operator. Our problem is to find a solution to the operator equation

$$\text{(2.1)} \qquad Px = \theta \; , \quad x \in X \; .$$

Consider the difference

$$P(x+h) - Px = Q(x)h$$

and suppose that $\Gamma(x)$ is a linear (additive and homogeneous) operator associated with $x \in X$ and acting from Y to X, i.e., $\Gamma(x):Y \to X$. Suppose that there exists an operator $A:E_Y \to E_Y$ such that $x+\Gamma(x) \in D$ (domain of P) implies

$$\text{(2.2)} \qquad \sigma(Q(x)\Gamma(x)y,y) \leq A\sigma(y,\theta) \; ,$$

where $x \in X$ and $y \in Y$ are determined by the problem. Condition (2.2) is equivalent to

(2.3) $$\sigma(P(x+\Gamma(x)y) - Px, y) \leq A\sigma(y,\theta) .$$

A is called a majorant operator. We say that A is positive if $\theta \leq \sigma \in E_Y$ implies $A\sigma \geq \theta$ and A is monotone if $\sigma \geq \sigma' \in E_Y$ implies $A\sigma \geq \sigma'$

Definition 2.1. We say that $\Gamma(x)$ is a contractor for P if there exists a positive monotone majorant operator A satisfying condition (2.2) or (2.3).

The contractor $\Gamma(x)$ is called bounded if there exists a positive linear continuous operator $B: E_Y \to E_X$ such that

(2.4) $$\rho(\Gamma(x)y,\theta) \leq B\sigma(y,\theta) \quad \text{for all} \quad y \in Y .$$

For solving equation (2.1) we use the following iteration procedure

(2.5) $$x_{n+1} = x_n - \Gamma(x_n)Px_n , \quad n = 0,1,\cdots ,$$

where x_0 is a given approximate solution to (2.1).

In order to investigate the convergence of this procedure, we assume that the majorant operator A satisfies the following condition:

(2.6) the series $a = \sum_{i=0}^{\infty} A^i\sigma(Px_0,\theta)$ is convergent.

Denote by X_0 the set of elements of X satisfying the inequality

(2.7) $$\rho(x_0-x,\theta) \leq Ba ,$$

where a is defined by (2.6).

Let $D \subset X$ be the domain of $P: D \to Y$. The operator P is said to be closed if

$$x_n \in D , \quad x_n \to x \quad \text{and} \quad Px_n \to y$$

imply $x \in D$ and $y = Px$.

The following theorem gives sufficient conditions for the convergence of the iterative procedure (2.5) to a solution of equation (2.1).

<u>Theorem 2.1</u>. Suppose that the nonlinear operator $P:D \to Y$ with $D \supset X_0$ is closed and has a contractor $\Gamma(x)$ satisfying inequality (2.3) for $x \in X_0$, i.e., satisfying (2.7) and provided that the majorant operator A satisfies condition (2.6). Then the sequence $\{x_n\}$ defined by (2.5) converges toward a solution x of equation (2.1). All x_n lie in X_0 and the error estimate holds:

$$\rho(x_n,x) \leq B \sum_{i=n}^{\infty} A^i \sigma(Px_0,\theta) \quad . \tag{2.8}$$

<u>Proof</u>. Putting $y = -Px_n$ in (2.3) we obtain

$$\sigma(Px_{n+1},\theta) \leq A\sigma(Px_n,\theta) \quad . \tag{2.9}$$

Since A is monotone, it follows from (2.9) that

$$\sigma(Px_n,\theta) \leq A^n\sigma(Px_0,\theta) \tag{2.10}$$

and, consequently, by (2.5) and (2.4),

$$\rho(x_n,x_{n+1}) \leq BA^n\sigma(Px_0,\theta) \quad . \tag{2.11}$$

By induction, it is easily seen that $x_n \in X_0$ and it follows from (2.11) that

$$\rho(x_n,x_m) \leq B \sum_{i=n}^{m-1} A^i \sigma(Px_0,\theta) \quad . \tag{2.12}$$

Inequality (2.10) implies $Px_n \to \theta$ as $n \to \infty$. By virtue of (2.12), $\{x_n\}$ is a Cauchy sequence and let x be its limit. Since P is closed, $x_n \to x$ and $Px_n \to \theta$ imply $Px = \theta$. The error estimate (2.8) results from (2.12). The inequality

$$\rho(x_0,x_n) \leq \sum_{i=0}^{n-1} A^i \sigma(Px_0,\theta)$$

is a particular case of (2.12) and implies $x \in X_0$.

In order to obtain a fixed point theorem let us consider the special case where

$$Px = x - Fx \quad \text{and} \quad Y = X \; .$$

The contractor inequality (2.3) yields here

$$\rho(Fx - F(x+\Gamma(x)y), (I - \Gamma(x))y) \leq A\rho(y,\theta) \quad (2.13)$$

After replacing (2.3) by (2.13), Theorem 2.1 remains true and yields a fixed point theorem for F.

In a special case where $\Gamma(x) \equiv I$ is the identity mapping of X, the contractor inequality (2.13) becomes

$$\rho(Fx, F(x+y)) \leq A\rho(y,\theta)$$

and we obtain a generalized contraction principle for F (see, for example, Krasnosel'skii and others [2]). In this case the iteration procedure (2.5) becomes exactly the method of successive approximation for F, i.e.,

$$x_{n+1} = Fx_n, \quad n = 0,1,\cdots.$$

Theorem 2.2. Suppose that, in addition to the hypotheses of Theorem 2.1, the contractor $\Gamma(x)$ is such that for arbitrary solutions $x, x' \in X_0$ of (2.1) the equation $x - \Gamma(x)y = x'$ has a solution for y. If for arbitrary $y \in Y$

$$A^n\sigma(y,\theta) \to \theta \text{ as } n \to \infty, \quad (2.14)$$

then equation (2.1) has a unique solution in X_0.

Proof. Since $Px = Px' = \theta$ by assumption, it follows from the contractor inequality (2.3) that

$$\sigma(y,\theta) \leq A\sigma(y,\theta) \leq \cdots \leq A^n\sigma(y,\theta)$$

for arbitrary positive integer n. By assumption, $A^n\sigma(y,\theta) \to 0$ as $n \to \infty$. Hence, $\sigma(y,\theta) = \theta$.

3. Contractors with monotone operator majorants

The contractor method will be used here with a different majorant operator A. Schröder [1] (see also Collatz [1]) has introduced this nonlinear majorant operator in his generalization of the method of successive approximations. This is a monotone nondecreasing operator which is also "difference-monotone". As far as linear pseudometric spaces are concerned, the generalized method of successive approximations is a very special case of the contractor method.

The notation and the problem are the same as in Section 2.

Suppose that the majorant operator $A:E_Y \to E_Y$ has the following "difference-monotone" property (see Collatz [1]):

$$\sigma,\eta,\sigma',\eta' \in E_Y \ , \ \theta \leq \sigma \leq \sigma' \quad \text{and} \quad \theta \leq \eta \leq \eta'$$

imply

$$\theta \leq A(\sigma+\eta) - A(\eta) \leq A(\sigma'+\eta') - A(\eta') \ . \tag{3.1}$$

In particular, for $\eta = \eta' = \theta$, it follows that

$$\theta \leq A(\sigma) \leq A(\sigma') \quad \text{for} \quad \theta \leq \sigma \leq \sigma' \ .$$

Let P be a nonlinear operator with domain $D \subset X$ and values in Y , where X and Y are linear pseudometric spaces.

<u>Definition 3.1</u>. We say that P has a contractor $\Gamma(x):Y \to X$ if there exists a positive monotone operator $A:E_Y \to E_Y$ satisfying condition (3.1) and such that $y,\bar{y} \in Y$ and $x+\Gamma(x)(y-\bar{y}) \in D$ imply

$$\sigma(P(x+\Gamma(x)(y-\bar{y}))-Px,y-\bar{y}) \leq A(\sigma(\bar{y},y)+\sigma(\bar{y},z)) - A\sigma(\bar{y},z) \tag{3.2}$$

for some fixed $z \in Y$.

To solve equation (2.1), we compare the iteration procedure (2.5) with the following one

$$\sigma_{n+1} = A\sigma_n + \gamma \ , \quad \gamma,\sigma_n \in E_Y \ , \quad n = 0,1,\cdots \ , \tag{3.3}$$

where γ is some fixed element. The following theorem gives sufficient conditions for the convergence of the iterative process (2.5) to a solution of equation (2.1).

<u>Theorem 3.1</u>. Suppose that

a) P is a closed nonlinear operator with domain $D \subset X$ and values in Y having a bounded linear contractor $\Gamma(x)$ which satisfies (2.4) and the contractor inequality (3.2) with positive monotone majorant A satisfying (3.1).

b) The sequence of σ_n defined by (3.3) converges to a solution $\sigma \in Y$ of equation $\sigma = A\sigma + \gamma$ and $x_0 \in X$, y_0, $z \in Y$, $\sigma_0 \in E_Y$, are chosen so as to satisfy

$$\sigma(Px_0,\theta) \leq \sigma_1 - \sigma_0 \quad \text{and} \quad \sigma(y_0,z) \leq \sigma_0 . \tag{3.4}$$

(In applications, one can choose $z = y_0$.)

c) The domain D contains the set K of elements $x \in X$ satisfying the inequality

$$\rho(x_0,x) \leq B(\sigma - \sigma_1) . \tag{3.5}$$

Then the sequence $\{x_n\}$ defined by (2.5) converges to a solution x of equation (2.1). All x_n and x belong to K and the error estimate holds

$$\rho(x,x_n) \leq B(\sigma - \sigma_n) , \quad n = 0,1,\cdots . \tag{3.6}$$

Proof. Since A is monotone, it follows from (3.4) and (3.3), by induction,

$$\sigma_{n+1} - \sigma_n = A\sigma_n - A\sigma_{n-1} \geq \theta$$

and using the properties of the convergence in metrizing space E_Y (see Collatz [1]), we obtain

$$\theta \leq \sigma_0 \leq \sigma_1 < \cdots \leq \sigma_n \leq \sigma . \tag{3.7}$$

Put $y_1 = y_0 - Px_0$. Then it follows from (2.5) and (3.2) with $x = x_0$, $y = y_1$, $\bar{y} = y_0$ that

$$\begin{aligned}\sigma(Px_1,\theta) &\leq A(\sigma(Px_0,\theta) + \sigma(y_0,z)) - A\sigma(y_0,z) \\ &\leq A((\sigma_1-\sigma_0) + \sigma_0) - A\sigma_0 = A\sigma_1 - A\sigma_0 = \sigma_2 - \sigma_1\end{aligned}$$

by virtue of (3.3), (3.4) and (3.1).

Thus,

$$\sigma(Px_1,\theta) \leq \sigma_2-\sigma_1 . \tag{3.8}$$

Further, putting $y_2 = y_1-Px_1$, we obtain from (2.5) and (3.2) with $x = x_1$, $y = y_2$ and $\bar{y} = y_1$ that

$$\sigma(Px_2,\theta) \leq A(\sigma(Px_1,\theta) + \sigma(y_1,z)) - A\sigma(y_1,z) .$$

But

$$\sigma(y_1,z) \leq \sigma(y_0,y_1) + \sigma(y_0,z) = \sigma(Px_0,\theta) + \sigma(y_0,z) \leq (\sigma_1-\sigma_0) + \sigma_0 = \sigma_1 ,$$

by (3.4).

Hence, by (3.1) and (3.8)

$$\sigma(Px_2,\theta) \leq A((\sigma_2-\sigma_1) + \sigma_1) - A\sigma_1 = A\sigma_2 - A\sigma_1 = \sigma_3 - \sigma_2 .$$

We prove by induction that

$$\sigma(Px_n,\theta) \leq \sigma_{n+1} - \sigma_n \quad \text{and} \quad \sigma(y_n,z) \leq \sigma_n , \tag{3.9$_n$}$$

where

$$y_{n+1} = y_n - Px_n .$$

In fact, it follows from (2.5) and (3.2) with $x = x_n$, $y = y_{n+1}$ and $\bar{y} = y_n$ that

$$\sigma(Px_{n+1},\theta) \leq A(\sigma(Px_n,\theta) + \sigma(y_n,z)) - A\sigma(y_n,z) \leq$$

$$\leq A((\sigma_{n+1}-\sigma_n) + \sigma_n) - A\sigma_n = \sigma_{n+2} - \sigma_{n+1} ,$$

by (3.9) and (3.1); now

$$\sigma_{n+1}(y_n,z) \leq \sigma(y_n,y_{n+1}) + \sigma(y_n,z) =$$

$$\sigma(Px_n,\theta) + \sigma(y_n,z) \leq (\sigma_{n+1}-\sigma_n) + \sigma_n ,$$

by (3.9) and (3.9$_{n+1}$) is true. It is easily seen that

$$\rho(x_n,x_m) \leq B(\sigma_m-\sigma_n) . \tag{3.10}$$

For we have, by (2.5), (2.4) and (3.9$_n$),

$$\rho(x_n,x_m) \le \sum_{i=n}^{m-1} \rho(x_i,x_{i+1}) = \sum_{i=n}^{m-1} \rho(\Gamma(x_i)Px_i,\theta) \le \sum_{i=n}^{m-1} B\sigma(Px_i,\theta) \le$$

$$\le \sum_{i=n}^{m-1} B(\sigma_{i+1}-\sigma_i) = B(\sigma_m-\sigma_n) .$$

In particular,

$$\rho(x_0,x_n) \le B(\sigma_n-\sigma_0) \le B(\sigma-\sigma_0) , \tag{3.11}$$

by (3.10) and (3.7).

We infer from b) and (3.9_n) that $Px_n \to \theta$ as $n \to \infty$. Inequality (3.10) proves that $\{x_n\}$ is a Cauchy sequence, that is, $\{x_n\}$ converges toward some element $x \in X$. But $x_n \in K \subset D$, by (3.11) and P is closed. Therefore $Px = 0$ and $x \in K$, by (3.11). The error estimate (3.6) results from (3.10) and the proof is completed.

Remark 3.1. Suppose that the majorant operator A is monotone and put $z = \theta \in Y$ in (3.2). Then (3.2) implies (2.3) provided that $A\theta = \theta$. If the contractor $\Gamma(x)$ satisfies (2.14) and for arbitrary solutions $x,x' \in D$ of (2.1) equation $x - \Gamma(x)y = x'$ has a solution for y , then (2.1) has at most one solution, by virtue of Theorem 2.2.

4. A fixed point theorem

A fixed point theorem can be obtained from Theorem 3.1 by putting $Px = x - Fx$, $Y = X$ and using the same argument as in section 2. The contractor inequality (3.2) yields here

$$\begin{aligned}&\rho(Fx-F(x+\Gamma(x)(y-\bar{y}) , (I-\Gamma(x))(y-\bar{y})) \le \\ &\le A(\rho(\bar{y},y) + \rho(\bar{y},z)) - A\rho(\bar{y},z) .\end{aligned} \tag{4.1}$$

Suppose now that $\Gamma(x) \equiv I$ is the identity mapping of X . Then (4.1) is equivalent to

$$\rho(Fx,Fy) \le A(\rho(x,y) + \rho(y,z)) - A\rho(y,z)$$

for arbitrary $x,y \in X$ and some fixed $z \in X$.

In this special case, we obtain a theorem by Schröder [1] (see also

Collatz [1]).

Thus, in linear pseudometric spaces Theorem 3.1 generalizes the theorem just mentioned.

It should be mentioned that the Banach contraction principle and its well-known generalizations quoted above require the domain and the range of the operator to lie in the same space while this restriction does not apply to the contractor method.

Another restriction seems to be also worthwhile to mention as far as the contraction principle is concerned. This restriction is related to the nature of the contraction mapping which must be continuous by necessity. Thus, in applications only continuous operators can be considered. However, this is not the case when using the contractor method. Here it is sufficient to assume that the operator in question is closed. This fact is usually important in applications.

APPENDIX 3

CONTRACTORS, APPROXIMATE IDENTITIES AND FACTORIZATION IN BANACH ALGEBRAS

Introduction. The concept of contractors is discussed in Chapters 1 - 3 as a tool for solving equations in Banach spaces. In this way, various existence theorems for solutions of equations are there obtained as well as convergence theorems for a broad class of iterative procedures. It is also shown there that the contractor method yields a unified approach to a large variety of iterative processes different in nature.

In this appendix, the contractor idea is introduced in Banach algebras.

Concerning the approximate identity in a Banach algebra A it is shown that if a subset U of A is a bounded left weak approximate identity, then U is a bounded left approximate identity. This important fact makes it possible to prove the well-known factorization theorems for Banach algebras under weaker conditions of existence of a bounded weak approximate identity.

A contractor is rather weaker than an approximate identity. Since every approximate identity is a contractor, the following seems to be a natural question: When is a contractor an approximate identity? The answer to this question is investigated in this appendix. The corresponding results are also applied to prove factorization theorems in Banach algebras (see Altman [14], [15], [16]).

2. Approximate identities

Let A be a Banach algebra.

Definition 2.1. A subset $U \subset B \subseteq A$ is called a left weak (or simple) approximate identity for the set B if for arbitrary $b \in B$ and $\epsilon > 0$ there exists an element $u \in U$ such that

(2.1) $$\|ub-b\| < \epsilon .$$

Definition 2.2. A subset $U \subset B \subseteq A$ is called a left approximate identity for B if for every arbitrary finite subset of elements $b_i \in B$ $(i = 1,2,\cdots,n)$ and arbitrary $\epsilon > 0$ there exists an element $u \in U$ such that

(2.2) $$\|ub_i-b_i\| < \epsilon \quad \text{for} \quad i = 1,2,\cdots,n .$$

A (weak) approximate identity U is called bounded if there is a constant d such that $\|u\| \leq d$ for all $u \in U$.

Lemma 2.1. If U is a bounded subset of $B \subseteq A$ such that for every pair of elements $b_i \in B$ $(i = 1,2)$ and arbitrary $\epsilon > 0$ there exists an element $u \in U$ satisfying (2.2) with $n = 2$, then U is a left approximate identity for B.

Proof. The proof will be given by the finite induction. Given arbitrary $b_i \in B$ $(i = 1,2,\cdots,n+1)$ and $\epsilon > 0$, for $\epsilon_0 > 0$ let $u_0 \in B$ be chosen so as to satisfy

(2.3) $$\|u_0b_i-b_i\| < \epsilon_0 \quad \text{for} \quad i = 1,2,\cdots,n \quad \text{and} \quad \|u_0\| \leq d .$$

For the pair u_0, $b_{n+1} \in B$ and $\epsilon_0 > 0$ there is an element $u \in U$ such that

(2.4) $$\|uu_0-u_0\| < \epsilon_0 \quad \text{and} \quad \|ub_{n+1}-b_{n+1}\| < \epsilon_0 , \ \|u\| \leq d .$$

After such a choice we have

$$\|ub_i-b_i\| \leq \|ub_i-uu_0b_i\| + \|uu_0b_i-u_0b_i\| + \|u_0b_i-b_i\| \leq$$
$$\leq d\epsilon_0 + M\epsilon_0 + \epsilon_0 < \epsilon \quad \text{for} \quad i = 1,2,\cdots,n$$

and

$$\|ub_{n+1}-b_{n+1}\| < \epsilon_0 < \epsilon ,$$

by (2.3) and (2.4), where $M = \max(\|b_i\|: i=1,2,\cdots,n)$ and $\epsilon_0 < (d+M+1)^{-1}\epsilon$.

Lemma 2.2. If the subset U of A is a bounded weak left approximate identity for $B \subseteq A$, then

$$U \circ U = [a \in A \mid a = u \circ v \,;\, u,v \in U] ,$$

where $u \circ v = u+v-uv$, has the following property: for every pair of elements $a,b \in B$ and $\epsilon > 0$ there exists $u \in U \circ U$ such that

$$\|ua-a\| < \epsilon \quad \text{and} \quad \|ub-b\| < \infty .$$

Proof. Given an arbitrary pair of elements $a,b \in B$ and $\epsilon > 0$, let $v \in U$ be chosen so as to satisfy

$$\|a-va\| < (1+d)^{-1}\epsilon \; , \; \|v\| \leq d . \tag{2.5}$$

For $b-vb$ and $\epsilon > 0$ there exists $w \in U$ such that

$$\|(b-vb) - w(b-vb)\| < \epsilon \; , \quad \|w\| \leq d .$$

Hence, we obtain

$$\|b-ub\| = \|b - (w+v-wv)b\| < \epsilon$$

and

$$\|a-ua\| = \|(a-va) - w(a-va)\| < (1+d)^{-1}\epsilon + d(1+d)^{-1}\epsilon = \epsilon ,$$

by (2.5), where $u = w+v-wv \in U \circ U$.

Lemma 2.3. If $U \subset A$ is a bounded weak left approximate identity for A , then U is a bounded left approximate identity for A .

Proof. By virtue of Lemma 2.2, the set $U \circ U$ satisfies the assumption of Lemma 2.1 and it can be replaced by U .

Remark 2.1. A partial result concerning this problem has been obtained by Reiter [1], §7, p. 30, Lemma 1: If A has a bounded left weak approximate identity, then A also has (multiple) left approximate identity (possibly unbounded).

<u>Theorem 2.1</u>. Let U be a bounded subset of the Banach algebra A satisfying the following conditions:

a) For every $u \in U \cup U \circ U$ and $\epsilon > 0$ there exists an element $v \in U$ such that $\|u-vu\| < \epsilon$.

b) For every element of the form $u-vu$ with $u,v \in U$ there exists an element $w \in U$ such that

$$\|(u-vu) - w(u-vu)\| < \epsilon .$$

Then $U \circ U$ is a bounded left approximate identity for the Banach algebra generated by U as well as for the right ideal generated by U . If U is commutative, then condition b) can be dropped.

<u>Proof</u>. Let $a,b \in U$ and $\epsilon > 0$ be arbitrary. By virtue of condition a), there exists $v \in U$ such that $\|a-va\| < (1+d)^{-1}\epsilon$, where d is the bound for U . Using b) for $b-vb$, we can choose $w \in U$ such that

$$\|(b-vb) - w(b-vb)\| < \epsilon .$$

Thus, we obtain

$$\|a-ua\| < \epsilon \quad \text{and} \quad \|b-ub\| < \epsilon ,$$

where $u = w+v-wv \in U \circ U$. Suppose that $b \in U \circ U$. Then for $\epsilon_0 > 0$ there exists $u_0 \in U$ such that $\|b-u_0b\| < \epsilon_0$. For $\epsilon > 0$ let $u \in U$ be chosen so as to satisfy

$$\|a-ua\| < \epsilon \quad \text{and} \quad \|u_0-uu_0\| < \epsilon .$$

Hence, we obtain

$$\|ub-b\| \leq \|ub-uu_0b\| + \|uu_0b-u_0b\| + \|u_0b-b\| \leq d\epsilon_0 + \|b\|\epsilon_0 + \epsilon_0 < \epsilon$$

for proper choice of ϵ_0 . If $a,b \in U \circ U$, then for $\epsilon_0 > 0$ choose $u_1,u_2 \in U$ such that

$$\|a-u_1a\| < \epsilon_0 \quad \text{and} \quad \|b-u_2b\| < \epsilon_0 .$$

Then we find $u \in U$ such that

$$\|u_1-uu_1\| < \epsilon_0 \quad \text{and} \quad \|u_2-uu_2\| < \epsilon_0 \ .$$

After such a choice we have

$$\|ua-a\| \leq \|ua-uu_1a\| + \|uu_1a - u_1a\| + \|u_1a - a\| \leq d\,\epsilon_0 + \|a\|\epsilon_0 + \epsilon_0 < \epsilon$$

for proper ϵ_0 , and similarly

$$\|ub-b\| \leq \|ub-uu_2b\| + \|uu_2b-u_2b\| + \|u_2b-b\| \leq d\,\epsilon_0 + \|b\|\epsilon_0 + \epsilon_0 < \epsilon$$

for proper ϵ_0 . Thus, by Lemma 2.1, $U \circ U$ is a bounded left approximate identity for $U \cup U \circ U$.

Now let $a = \sum_{i=1}^{n} u_i a_i$ and $b = \sum_{j=1}^{m} v_j b_j$, where $u_i, v_j \in U$ and $a_i, b_j \in A$ for $i = 1,\cdots,n$; $j = 1,\cdots,m$. For $\epsilon_0 > 0$ choose $u \in U \circ U$ such that $\|u_i-uu_i\| < \epsilon_0$ and $\|v_j-uv_j\| < \epsilon_0$ for $i = 1,\cdots,n$ and $j = 1,\cdots,m$. Then

$$\|a-ua\| \leq \|\sum_{i=1}^{n} (u_i-uu_i)a_i\| < \epsilon_0 \sum_{i=1}^{n} \|a_i\| < \epsilon$$

for sufficiently small ϵ_0 . The same holds for b , that is $\|b-ub\| < \epsilon$. The assertion of the theorem follows now from Lemma 2.1. If U is commutative, then b) follows from a). For let $a = u-vu$, $u,v \in U$. Then

$$\|a-wa\| = \|(u-ww) - (u-wu)v\| < \epsilon$$

if $w \in U$ is such that $\|u-wu\| < (1+d)^{-1}\epsilon$.

For the set $U \subset A$ let us define an infinite sequence of sets $\{P_n\}$ as follows. Put $P_1 = U$, $P_2 = U \circ U$. Then $P_n = U \circ P_{n-1} = U \circ U \circ \ldots \circ U$ (n times) is the set of all elements p of the form $p = u+v-uv$, where $u \in U$ and $v \in P_{n-1}$. Let P be the union of all sets P_n , that is $P = P_1 \cup P_2 \cup \ldots$.

<u>Lemma 2.4</u>. If U is a left bounded approximate identity for itself, then U is the same for the Banach algebra generated by U and in particular for P .

<u>Proof</u>. The proof follows from the argument used at the end of the proof

of Theorem 2.1.

3. Contractors in Banach algebras

Definition 3.1. A subset U of a Banach algebra A is called a left contractor for A if there is a positive constant $q < 1$ with the following property.

For every $a \in A$ there exists an element $u \in U$ (depending on a) such that

$$\|a-ua\| \leq q\|a\| . \tag{3.1}$$

A contractor U is said to be bounded if U is bounded by some constant d.

Lemma 3.1. Let U be a left contractor for A. Then for arbitrary $a \in A$, there exists an infinite sequence $\{a_n\} \subset P$ such that

$$\|a-a_n a\| \leq q^n\|a\| \quad \text{and} \quad a_n \in P_n . \tag{3.2}$$

Proof. By (3.1), let $u_1 \in U$ be chosen so as to satisfy the inequality

$$\|a-u_1 a\| \leq q\|a\| . \tag{3.3}$$

Now let $u_2 \in U$ be such that

$$\|(a-u_1 a) - u_2(a-u_1 a)\| \leq q\|a-u_1 a\| . \tag{3.4}$$

Hence, we obtain from (3.3) and (3.4)

$$\|a-a_2 a\| \leq q^2\|a\| , \quad \text{where} \quad a_2 = u_2+u_1-u_2u_1 = u_2 \circ u_1 \in P_2 .$$

We repeat this procedure replacing in (3.4) u_1 by a_2 and u_2 by u_3. Thus, we have $a_3 = u_3 \circ a_2 \in P_3$. After n iteration steps we obtain (3.2).

Definition 3.2. A subset $U \subset A$ is called a strong left contractor for A if there exists a positive $q < 1$ with the following property: For every arbitrary finite set of $a_i \in A$, $i = 1,2,\cdots,n$, there is an element $u \in U$ such that

(3.5) $$\|a_i - ua_i\| \leq q\|a_i\| , \quad i = 1,2,\cdots,n .$$

A left contractor U is said to be quasi-strong if for arbitrary pair of $a_i \in A$ $(i = 1,2,)$ there exists an element $u \in U$ satisfying (3.5) with $n = 2$.

Lemma 3.2. Let $U \subset A$ be a left quasi-strong contractor for A . Then for every arbitrary pair of $a,b \in A$, there exists an infinite sequence $\{c_n\} \subset P$ such that

(3.6) $$\|a - c_n a\| \leq q^n\|a\|, \ \|b - c_n b\| \leq q^n\|b\|$$

where $c_n \in P_n$, $n = 1,2,\cdots$.

Proof. The proof is similar to that of Lemma 3.1.

A similar lemma holds for strong contractors.

Lemma 3.3. Let $U \subset A$ be a left strong contractor for A . Then for every arbitrary finite set of elements $a_i \in A$, $i = 1,2,\cdots,m$, there exists an infinite sequence $\{c_n\} \subset P$ such that

$$\|a_i - c_n a_i\| \leq q^n\|a_i\| \quad \text{for} \quad i = 1,2,\cdots,m ,$$

where $c_n \in P_n$, $n = 1,2,\cdots$.

Lemma 3.4. Suppose that U is a left bounded contractor for A satisfying the condition $(d+1)q < 1$. Then $U \circ U$ is a left bounded quasi-strong contractor for A .

Proof. Let $\bar{q} = (d+1)q < 1$ and let $a,b \in A$ be arbitrary. Then choose $v \in U$ so as to satisfy

$$\|a - va\| \leq q\|a\|, \ \|v\| \leq d .$$

For $b-vb$ let $w \in U$ be such that

$$\|(b-vb) - w(b-vb)\| \leq q\|b-vb\| .$$

Put $u = w + v - Wv \in U \circ U$. Then

$$\|a-ua\| = \|(a-va) - w(a-va)\| \leq q\|a\| + dq\|a\| = \bar{q}\|a\|$$

and

$$\|b-ub\| = \|(b-vb) - w(b-vb)\| \leq q\|b-vb\| \leq q\|b\| + dq\|b\| = \bar{q}\|b\| .$$

Thus, $U \circ U$ is a bounded left quasi-strong contractor for A with contractor constant $\bar{q} < 1$.

Theorem 3.1. A left bounded contractor U for A is a left bounded approximate identity for A if and only if U is a left approximate identity for itself.

Proof. Let $a,b \in A$ and $\epsilon > 0$ be arbitrary. Using Lemma 3.1, we construct a sequence $\{a_n\} \subset P$ for $a \in A$ and $\{b_n\} \subset P$ for $b \in A$ such that

$$\|a-a_n a\| \leq q^n\|a\| \quad \text{and} \quad \|b-b_n b\| \leq q^n\|b\| , \tag{3.7}$$

where $a_n, b_n \in P_n$, $i = 1,2,\cdots$. By virtue of Lemma 2.4, for $a_n, b_n \in P_n \subset P$ and $\epsilon_0 > 0$ we can choose $u \in U$ so as to satisfy

$$\|a_n - ua_n\| < \epsilon_0 \quad \text{and} \quad \|b_n - ub_n\| < \epsilon_0 .$$

Then we obtain, by (3.7),

$$\|ua-a\| \leq \|ua-ua_n a\| + \|ua_n a-a_n a\| + \|a_n a-a\| < dq^n\|a\| + \epsilon_0\|a\| + q^n\|a\| < \epsilon$$

for sufficiently large n and proper choice of ϵ_0 . A similar estimate holds for b :

$$\|ub-b\| \leq dq^n\|b\| + \epsilon_0\|b\| + q^n\|b\| < \epsilon$$

for sufficiently large n and proper choice of ϵ_0 . The proof of necessity is obvious.

Theorem 3.2. Let U be a bounded left contractor for A . If U satisfies the hypotheses of Theorem 2.1, then $U \circ U$ is a left bounded approximate identity for A .

Proof. The proof is the same as that of Theorem 3.1. The only difference is replacing there Lemma 2.4 by Theorem 2.1.

<u>Theorem 3.3</u>. Let U be a left bounded contractor for A satisfying the condition $(d+1)^3 q < 1$. If U is a weak left approximate identity for $U \circ U$, then U is a bounded left approximate identity for A .

<u>Proof</u>. Let q and $\epsilon_0 > 0$ be such that

$$(d+1)^3 q < ((d+1)^3 + 2\epsilon_0) q \leq \bar{q} < 1 .$$

By Lemma 3.4, $U \circ U$ is a quasi-strong contractor for A with contractor constant $(d+1)q$. Hence, for arbitrary $a, b \in A$ and $u_1 \in U$ there exists an element $w \in U \circ U$ such that

$$\|(a-u_1 a) - w(a-u_1 a)\| \leq (d+1)q\|a-u_1 a\|$$

and

$$\|(b-u_1 b) - w(b-u_1 b)\| \leq (d+1)q\|b-u_1 b\| .$$

By assumption, there exists $v \in U$ such that $\|w-vw\| < \epsilon_0 q(d+1)^{-1}$. Therefore,

$$\|v(a-u_1 a) - (a-u_1 a)\| \leq \|v(a-u_1 a) - vw(a-u_1 a)\| + \|vw(a-u_1 a) - w(a-u_1 a)\| +$$

$$+ \|w(a-u_1 a) - (a-u_1 a)\| < (d(d+1)q + \epsilon_0 (d+1)^{-1} q + (d+1)q) \|a-u_1 a\| .$$

Hence,

$$\|a-(v \circ u_1)a\| \leq ((d+1)^2 q + \epsilon_0 q(d+1)^{-1}) \|a-u_1 a\| . \tag{3.8}$$

Using the assumption again we can find $u_2 \in U$ such that

$$\|(v \circ u_1) - u_2(v \circ u_1)\| < \|a\|^{-1} \epsilon_0 q\|a-u_1 a\| . \tag{3.9}$$

Hence, we have, by (3.8) and (3.9),

$$\|u_2 a-a\| \leq \|u_2 a-u_2(v \circ u_1)a\| + \|u_2(v \circ u_1)a - (v \circ u_1)a\| + \|(v \circ u_1)a-a\| \leq$$

$$\leq [d((d+1)^2 q + \epsilon_0 (d+1)^{-1} q) + q\epsilon_0 + ((d+1)^2 q + \epsilon_0 (d+1)^{-1} q)] \|a-u_1 a\| .$$

Thus, we obtain,

$$\|a-u_2 a\| \leq \bar{q}\|a-u_1 a\| , \quad u_2 \in U . \tag{3.10}$$

Similarly, we get

(3.11) $$\|b-u_2b\| \leq \bar{q}\|b-u_1b\| .$$

Since $u_1 \in U$ was arbitrary, by the same argument, for $u_2 \in U$, there exists $u_3 \in U$ satisfying conditions (3.10) and (3.11) with u_2 and u_3 replacing u_1 and u_2, respectively. After $n-1$ iteration steps we obtain

$$\|a-u_na\| \leq \bar{q}^n\|a-u_1a\| < \epsilon$$

and

$$\|b-u_nb\| \leq \bar{q}^n\|b-u_1b\| < \epsilon , \quad u_n \in U ,$$

for sufficiently large n. Since a,b and $\epsilon > 0$ are arbitrary, it follows from Lemma 2.1 that U is a bounded left approximate identity for A.

Using the same technique, one can prove the following

Proposition 3.1. A left bounded quasi-strong contractor U for A is a left approximate identity for A if and only if U is a left weak approximate identity for an infinite subsequence of $\{P_n\}$.

Proof. Let $a,b \in A$ and $\epsilon > 0$ be arbitrary. Using Lemma 3.2 for the pair a,b we construct an infinite sequence $\{c_n\}$ satisfying (3.6). Now let us choose $u \in U$ so as to satisfy $\|uc_n-c_n\| < \epsilon_0$ for infinitely many n. Then we obtain for $a \in A$

$$\|ua-a\| \leq \|ua-uc_na\| + \|uc_na-c_na\| + \|c_na-a\| < dq^n\|a\| + \epsilon_0\|a\| + q^n\|a\| < \epsilon$$

for some sufficiently large n and proper choice of ϵ_0. A similar estimate holds for b:

$$\|ub-b\| < dq^n\|b\| + \epsilon_0\|b\| + q^n\|b\| < \epsilon$$

for some sufficiently large n and proper choice of ϵ_0. The proof of necessity is obvious.

As an immediate corollary to Proposition 3.1, we obtain the following

Proposition 3.2. A left bounded weak approximate identity U for U is a left approximate identity for A if and only if U is a left quasi-strong contractor for A.

Proposition 3.3. Suppose that U is a left bounded contractor for A satisfying the condition $(d+1)q < 1$. Then U is a left bounded weak approximate identity for A if and only if U is the same for $U \circ U$.

Proof. Let $\bar{q}$ and ϵ_0 be such that

$$(d+1)q < (d+1+\epsilon_0)q \leq \bar{q} < 1 . \tag{3.12}$$

For arbitrary $a \in A$, let $u_1 \in U$ be such that $\|u_1 a-a\| \leq q\|a\|$. Then choose $v_1 \in U$ so as to satisfy

$$\|v_1(u_1 a-a) - (u_1 a-a)\| \leq q\|u_1 a-a\| ,$$

or equivalently, $\|a_2 a-a\| \leq q\|u_1 a-a\|$ with $a_2 = v_1 \circ u_1 \in U \circ U$. By assumption, for a_2 there is an element $u_2 \in U$ such that

$$\|u_2 a_2-a_2\| < \epsilon_0\|a_2 a-a\| \cdot \|a\|^{-1} .$$

Hence,

$$\|u_2 a-a\| \leq \|u_2 a_2-u_2 a_2 a\| + \|u_2 a_2 a-a_2 a\| + \|a_2 a-a\| < dq\|u_1 a-a\| +$$

$$+ \epsilon_0\|a_2 a-a\| + q\|u_1 a-a\| \leq (d+1+\epsilon_0)q\|u_1 a-a\| \leq \bar{q}\|u_1 a-a\| ,$$

by (3.12). Thus, for arbitrary $u_1 \in U$ there exists an element $u_2 \in U$ such that

$$\|u_2 a-a\| \leq \bar{q}\|u_1 a-a\| .$$

After n iteration steps we obtain

$$\|u_n a-a\| \leq \bar{q}^n\|u_1 a-a\| < \epsilon \quad (u_n \in U)$$

if n is sufficiently large.

4. Factorization theorems

Let A be a Banach algebra and let X be a Banach space. Suppose that there is a composition mapping of $A \times X$ with values $a \cdot x$ in X.

X is called a left Banach A-module (see Hewitt and Ross [2], II (32.14)), if this mapping has the following properties:

(i) $(a+b)\cdot x = a\cdot x+b\cdot x$ and $a\cdot(x+y) = a\cdot x+a\cdot y$;

(ii) $(ta)\cdot x = t(a\cdot x) = a\cdot(tx)$;

(iii) $(ab)\cdot x = a\cdot(b\cdot x)$;

(iv) $\|a\cdot x\| \leq C\|a\|\cdot\|x\|$

for all $a,b \in A$; $x,y \in X$; real or complex t , where C is a constant ≥ 1 . Denote by A_e the Banach algebra obtained from A by adjoining a unit e , and with the customary norm $\|a+te\| = \|a\| + |t|$. Properties (i) - (iv) hold for the extended operation $(a+te)\cdot x = a\cdot x+tx$.

The well-known factorization theorems for Banach algebras and their extension to Banach A-modules are usually proved under the hypothesis that the Banach algebra A has a bounded (left) approximate identity in the sense defined by Definition 2.2. Thus, it follows from Lemma 2.3 that all factorization theorems in question remain true under the weaker assumption of the existence of a bounded weak left approximate identity for A .

Let U be a bounded weak left approximate identity for A . Put $W = U\circ U$ and denote by d the bound for W .

<u>Theorem 4.1</u>. Let A be a Banach algebra having a bounded weak left approximate identity U . If X is a left Banach A-module, then $A\cdot X$ is a closed linear subspace of X . For arbitrary $z \in A\cdot X$ and $r > 0$ there exists an element $a \in A$ and an element $x \in X$ such that $z = a\cdot x$, $\|z-x\| \leq r$, where x is in the closure of $A\cdot z$.

<u>Proof</u>. It is easy to see that if z is in the closure of $A\cdot X$, then for arbitrary $a \in A$ and $\epsilon > 0$ there exists $u \in W$ such that

$$\|ua-a\| < \epsilon \quad \text{and} \quad \|u\cdot z-z\| < \epsilon . \tag{4.1}$$

In fact, for $\epsilon_0 > 0$ there exist $b \in A$ and $y \in X$ such that $\|b\cdot y-z\| < \epsilon_0$. Since U is a weak bounded left approximate identity for A, by Lemma 2.2, for $\epsilon_0 > 0$ there exists $u \in W$ such that

$$\|ua-a\| < \epsilon_0 \quad \text{and} \quad \|ub-b\| < \epsilon_0 .$$

Hence, we obtain

$$\|u\cdot z-z\| \leq \|u\cdot z-ub\cdot y\| + \|ub\cdot y-b\cdot y\| + \|b\cdot y-z\| \leq dC\|z-b\cdot y\| +$$
$$+ \epsilon_0 C\|y\| + \epsilon_0 < (dC+C\|y\|+1)\epsilon_0 < \epsilon$$

for sufficiently small ϵ_0. Now put

$$a_0 = e, a_{n+1} = (2d+1)^{-1}(u_{n+1}+2de)a_n \ ; \quad n = 1,2,\cdots .$$

We have

$$a_n = a'_n + q^n e \ , \quad \text{where} \quad a'_n \in A \ , \quad q = 2d(2d+1)^{-1} \ ;$$

$$a_{n+1}-a_n = (2d+1)^{-1}(u_{n+1}a_n-a_n) = (2d+1)^{-1}(u_{n+1}a'_n-a'_n) + (2d+1)^{-1}q^n(u_{n+1}-e);$$

$$a_{n+1}^{-1}-a_n^{-1} = a_n^{-1}(2d+1)(u_{n+1}+2de)^{-1}-a_n^{-1} = a_{n+1}^{-1}(e-(2d+1)^{-1}(u_{n+1}+2de)) =$$
$$= (2d+1)^{-1}a_{n+1}^{-1}(e-u_{n+1}) ,$$

where $a_n^{-1} \in A_e$. Let $x_n = a_n^{-1}\cdot z$. Then we obtain

$$\|x_{n+1}-x_n\| \leq C(2d+1)^{-1}\|a_{n+1}^{-1}\|\|z-u_{n+1}\cdot z\| .$$

Since $\|a_n^{-1}\| \leq (2+d^{-1})^n$, let us choose u_{n+1} so as to satisfy (4.1) with $a = a'_n$ and $\epsilon = \epsilon_n = C^{-1}(2d+1)(2+d^{-1})^{-1-n}2^{-1-n}r$. Hence, we have

$$\|u_{n+1}a'_n-a'_n\| < \epsilon_n \quad \text{and} \quad \|x_{n+1}-x_n\| \leq 2^{-1-n}r .$$

It follows that the sequences $\{a_n\}$ and $\{x_n\}$ converge toward $a \in A$ and $x \in X$, respectively. Evidently,

$$z = a\cdot x \quad \text{and} \quad \|z-x\| \leq \sum_{n=0}^{\infty}\|x_{n+1}-x_n\| \leq r .$$

By (4.1), z is in the closure of $A\cdot z$ and so are $x_n = a^{-1}z$, and, consequently, so is x . Thus, $A \cdot X$ is closed and its linearity follows from the following observation. For arbitrary $a,b \in A$; $x,y \in X$ and $\epsilon > 0$ let $u \in W$ be such that

$$\|ua-ua\| < C^{-1}(\|x\|+\|y\|)^{-1} \quad \text{and} \quad \|ub-b\| < C^{-1}(\|x\|+\|y\|)^{-1}\epsilon \ .$$

Then we have

$$\|a \cdot x+b \cdot y-u(a \cdot x+b \cdot y)\| = \|(a-ua) \cdot x + (b-ub) \cdot y\| < C\|a-uax\|\|x\| + C\|b-ub\|\|y\| < \epsilon \ .$$

That is, $a \cdot x+b \cdot y$ is in the closure of $A \cdot X$.

Remark 4.1. Theorem 4.1 generalizes the factorization theorems of Cohen [1], Hewitt [1], Curtis and Figa-Talamanca [1] (Collins and Summers [1], Hewitt and Ross [2]: (32.22), (32.23), (32.26), Altman [14], [15], [16]. For more facts and references concerning approximate identities and factorization, see Doran and Wichmann [1]).

In terms of contractors, Theorem 4.1 can be formulated as

Theorem 4.2. Suppose that the Banach algebra A has a left bounded (by d) contractor U satisfying one of the following conditions

a) U is a left approximate identity for itself.

b) U satisfies the hypotheses of Theorem 2.1.

c) $(d+1)q < 1$ and U is a weak left approximate identity for $U \circ U$.

Then all assertions of Theorem 4.1 hold.

A corollary to Theorem 4.1 is the following generalization of the well-known theorem (see Hewitt and Ross [2], II(32.23)).

Theorem 4.3. Let A be a Banach algebra with a weak bounded left approximate identity U . Let $\zeta = \{z_n\}$ be a convergent sequence of elements of $A \cdot X$, and suppose that $r > 0$. Then there exists an element $a \in A$ and a convergent sequence $\xi = \{x_n\}$ of elements of $A \cdot X$ such that

$$z_n = a \cdot x_n \quad \text{and} \quad \|z_n - x_n\| \leq r \quad \text{for} \quad n = 1, 2, \cdots ,$$

where x_n is in the closure of $A \cdot z_n$.

Proof. Let $\mathcal{X}$ be the Banach space of all convergent sequences $\xi = \{x_n\}$ of elements of the closed linear subspace $A \cdot X$ of X with the norm $\|\xi\| = \sup(\|x_n\| : n = 1, 2, \cdots)$. Consider the left Banach A-module $\mathcal{X}$ with $a \cdot \xi = \{a \cdot x_n\} \in \mathcal{X}$. For $\xi \in \mathcal{X}$ put $\xi_m = \{x_n\} \in \mathcal{X}$ with $x_n = x_m$ for $n \geq m$. By Theorem 4.1, it is sufficient to show that every $\xi \in \mathcal{X}$ is in the closure of $A \cdot \mathcal{X}$. But $\xi_m \to \xi$ as $m \to \infty$. Therefore let $\xi_m = \{a_n \cdot x_n\} \in \mathcal{X}$ with $a_n \cdot x_n = a_m \cdot x_m$ for $n \geq m$. By Lemma 2.3, for $\epsilon_0 > 0$ there exists $u \in A$ such that

$$\|ua_i - a_i\| < \epsilon_0 \quad \text{for} \quad i = 1, \cdots, m .$$

Hence, we have

$$\|ua_i \cdot x_i - a_i \cdot x_i\| < c\epsilon_0 \|x_i\| < \epsilon$$

for sufficiently small ϵ_0 and, consequently, $\|u \cdot \xi_m - \xi_m\| < \epsilon$, where $\epsilon > 0$ is arbitrary.

Remark 4.1. In Theorem 4.3, convergent sequences can be replaced by sequences convergent toward zero. Then $\mathcal{X}$ will be the space of all sequences of $A \cdot X$ convergent to zero.

5. The converse of Cohen's factorization theorem

Definition 5.1. A Banach algebra A possesses the Cohen factorization property if there exists a constant number d such that, for arbitrary $z \in A$ and $r > 0$, there are elements $a, x \in A$ which satisfy the following conditions:

(i) $z = ax$

(ii) $\|x - z\| < r$ and

(iii) $\|a\| \leq d$.

Lemma 5.1. If A possesses the Cohen factorization property, then A has a bounded left approximate identity.

Proof. Let $\{r_n\}$ be a sequence of positive numbers converging to 0. Let z be an arbitrary element of A. Then there exist elements a_n and x_n in A such that

$$z = a_n x_n, \quad \|x_n - z\| < r_n \quad \text{and} \quad \|a_n\| \leq d .$$

Hence, we obtain

$$\|z - a_n z\| = \|a_n x_n - a_n z\| \leq \|a_n\| \cdot \|x_n - z\| \to 0 \quad \text{as} \quad n \to \infty .$$

Thus, the set U of elements of all $\{a_n\}$ chosen for every z of A is a bounded left weak approximate identity for A. By virtue of Lemma 2.3, the set U is a bounded left approximate identity for A.

Theorem 5.1. A Banach algebra has the Cohen factorization property if and only if it has a simple or multiple bounded left (right) approximate identity.

Proof. The proof follows from the Cohen factorization theorem and from Lemma 5.1.

6. A necessary and sufficient condition for a Banach module to be essential

Definition 6.1. Let A be a Banach algebra and let X be a Banach space. A Banach left A-module X is called essential if $A \cdot X = X$.

Lemma 6.1. Let A be a Banach algebra with a left bounded approximate identity and let y be an element of the Banach left A-module X. Then $y \in A \cdot X$ if and only if there exists a positive constant $q < 1$ which has the following property: (α) For every pair $a \in A$, $x \in X$ there exists a pair $\bar{a} \in A$, $\bar{x} \in X$ and a positive number $\epsilon \leq 1$ such that

$$\|\bar{a}\bar{x} - ax - \epsilon(y - ax)\| \leq q\epsilon\|y - ax\| . \tag{6.1}$$

Proof. The proof follows immediately from Theorem 1.1, Chapter 10, since the set $Z = A \cdot X$ is closed in X, by virtue of Theorem 4.1.

On the other hand, condition (6.1) means that the set $\Gamma_z(Z)$ of contractor directions for Z contains the elements $y-z$ for all $z = ax \in Z$ and the proof is complete.

As a consequence of Lemma 6.1, we obtain the following

Theorem 6.1. Suppose that the Banach algebra A has a left approximate identity. Then the Banach left A-module X is essential if and only if for arbitrary $y \in X$, there exists a positive constant $q = q(y) < 1$ which has the property (α).

Proof. The proof follows immediately from Lemma 6.1.

Now we do not assume that the Banach algebra A has an approximate identity. In this case we can only obtain sufficient conditions.

Lemma 6.2. Let y be an element of X. Suppose that there exists a positive $q < 1$ and a continuous increasing function $B(s) > 0$ for $s > 0$ with

$$\int_0^1 s^{-1} B(s)\,ds < \infty \tag{6.2}$$

which have the following property: (β) For every pair $a \in A$, $x \in X$, there exists a pair $\bar{a} \in A$, $\bar{x} \in X$ and a positive number $\epsilon \leq 1$ which satisfy condition (6.1) and

$$\|\bar{a}-a\| \leq B(\|y-ax\|) \quad \text{and} \quad \|\bar{x}-x\| \leq B(\|y-ax\|) \,. \tag{6.3}$$

Then $y \in A \cdot X$.

Proof. The proof uses the method of contractor directions and is similar to that of Theorem 4.2, where the constant B can be replaced by the function $B(s)$ determined in property (β). Let us mention that the proof requires the transfinite induction to be applied to both sequences $\{a_\alpha\} \subset A$ and $\{x_\alpha\} \subset X$ as well as to the sequence $\{y_\alpha - a_\alpha x_\alpha\}$.

By virtue of Lemma 6.2, we obtain the following

Theorem 6.2. Suppose that for arbitrary $y \in X$, there exist a positive constant $q = q(y) < 1$ and an increasing continuous function $B(s) > 0$ for $s > 0$ with condition (6.2), which have the property (β). Then $A \cdot X = X$.

Proof. The proof follows immediately from Lemma 6.2.

7. The general factorization property and unsolved problems

Definition 7.1. A Banach algebra A is said to have the factorization property if for arbitrary $z \in A$, there exist elements $z, x \in A$ such that $z = ax$.

It follows from Definition 7.1 that a Banach algebra A has the (general) factorization property if and only if $A \cdot A = A$, i.e., the Banach algebra A considered as a left or right Banach A-module is essential. Hence, it follows that the general factorization problem for a Banach algebra A is a particular case of a more general problem of essentiality of a Banach left or right A-module. Let us notice that Theorem 6.1 is not applicable in this case, since we do not assume that Z is closed in A, where $Z = A \cdot A \subset A$. Therefore, some clues to the general factorization problem of a Banach algebra can be given by Theorem 6.2. Although condition (6.1) is certainly necessary for the factorization property, condition (6.3) appears to be only a sufficient one. Under additional hypotheses, condition (6.3) can be replaced by a more general one of the type (1.2), Chapter 6, by exploiting a technique similar to that used in Section 1, Chapter 6.

Moreover, we can consider the factorization property for an individual element of a Banach algebra. In this case, we can exploit a aimilar technique as used for local existence theorems proved in Section 1, Chapter 6.

It is known that the factorization property can be valid for a Banach algebra which does not possess an approximate identity. In this context the following seem to be natural questions:

A) Suppose that the Banach algebra A has a bounded left contractor.

Does the Banach algebra A possess the factorization property?

A^*) Suppose that the Banach algebra A has a bounded left contractor. Does the Banach algebra A possess a bounded left approximate identity?

An affirmative answer to the question A^*) yields automatically an affirmative answer to question A), by virtue of the Cohen factorization theorem.

CONCLUDING REMARKS

Let us conclude with a few remarks. First of all this work can be viewed basically as a development of a non-differential calculus for iterative and semiiterative methods of solving general equations. This calculus rests upon two fundamental notions: contractors and contractor directions. The contractor concept appears to be flixible and is easily combined with the notion of majorant functions. The concept of a majorant function introduced by Kantorovič was also used by other authors. Let us mention that in our case the role and character of the majorant functions are rather different and can be used not only for iterative procedures but also in order to investigate the structure of the operator itself independently. As far as the arithmetics involved here are concerned we have exploited and further developed the ones known in the literature, for instance, the method used by Ortega and Rheinboldt. Moreover, we have also developed entirely new techniques, for instance, the method of series of iterates of a positive function. In passing by, let us mention that series of iterates constitute a new branch of independent interest in the theory of infinite series of positive terms. The unified theory of iterative methods presented in this monograph contains as particular cases a large number of iteration procedures including the most important ones which are known in the literature. However, it was not our intention to make a review of all known methods which can be treated from our standpoint.

We have seen that the theory of iterative methods which applies to approximate solutions of operator equations is not applicable to nonlinear functionals. Therefore, we have shown that it is possible to develop a parallel theory for finding approximate roots of nonlinear functionals on a Banach space. Moreover, even the arithmetic used here is similar. This fact is due to the introduction of independent concepts of contractors and two point contractors for nonlinear functionals. As far as applications are concerned, we merely mentioned Yoshimura who gave an application of the contractor method with nonlinear majorants to the Quantum Field Theory. But we introduced the contractor idea in Banach algebras investigating approximate identities and putting some new questions.

In order to make useful the concept of contractor directions we adopted and considerably developed the transfinite induction technique used by Gavurin. In this way, we ultimately obtained a new very general existence principle in nonlinear functional analysis (see Section 1, Chapter 6).[†] We have shown how to apply this principle to solutions of some problems in nonlinear partial differential equations and nonlinear integral equations as well. However, in order to widen the scope of applicability of the new principle, extensive studies of solutions of linearized equations in both fields are needed. Finally, we hope that this work will open new directions and challenges in both theory and applications.

[†] For further results see: M. Altman, An existence principle in nonlinear functional analysis, to appear.
Iterative methods which are based on the concept of contractor directions are investigated in: M. Altman, A strategy theory of solving equations, to appear.

REFERENCES

M. Altman

1. Inverse differentiability, contractors and equations in Banach spaces, Studia Math. 46(1973), 1-15.

2. Directional contractors and equations in Banach spaces, ibid., 101-110.

3. Contractors with nonlinear majorant functions and equations in Banach spaces, Bollet. Un. Mat. Ital. (4) 9 (1974), 615-637.

4. Contractors and equations in pseudometric spaces, ibid., (4) 6 (1972), 376-384.

5. The contractor theory of solving equations, in: Numerical solutions of systems of nonlinear equations, 373-414. Acad. Press, 1973.

6. A class of majorant functions for contractors and equations, Bull. Austral. Math. Soc. 10(1974), 51-58.

7. An integral test and generalized contractions, Amer. Math. Monthly, 82(1975), 827-829.

8. Contractors and secant methods for finding roots of nonlinear functionals, J. Austral. Math. Soc. 19(1975), 74-90.

9. Solvability of nonlinear operator equations, to appear.

10. Contractor directions, directional contractors and directional contractions for solving equations, Pacif. J. Math., Vol. 62, No. 1(1976).

11. The method of contractor directions for partial differential equations, to appear.

12. Series of iterates, to appear.

13. General solvability theorems, to appear.

14. Infinite products and factorization in Banach algebras, Boll. Un. Mat. Ital. (4) 5 (1972), 217-229.

15. Contractors, approximate identities and factorization in Banach algebras, Pacif. J. Math. 48(1973), 323-334.

16. A generalization and the converse of Cohen's factorization theorem, Duke Math. J., vol. 42(1975), 105-110.

17. A generalization of Newton's method, Bull. Acad. Polon. Sci., Cl. III, 3(1955), 189-193.

Altman (cont.)

18. Concerning approximate solutions of nonlinear functional equations, ibid., vol. 5(1957), 461-465.

19. On the approximate solutions of operator equations in Hilbert space, ibid., vol. 5(1957), 605-609.

20. Connection between gradient methods and Newton's method for functionals, ibid., vol. 9(1961), 877-880.

21. On the approximate solutions of functional equations in L^pspaces, Colloquium Math., vol. 6(1958), 127-134.

R. G. Bartle

1. Newton's method in Banach spaces, Proc. Amer. Math. Soc. Vol. 6 (1955), 827-831.

B. A. Bel'tyukov

1. On a perturbed analog of the method of Aitken-Steffensen for the solution of nonlinear operator equations, Sibirskii Mat. Zhurnal, Vol. 12(1971), No. 5, 983-1000.

A. Bielecki

1. Une remarque sur la méthode de Banach-Cacciopoli-Tikhonov dans la théorie des équations différentielles ordinaires, Bull. Acad. Polon. Sci. Cl. III, 4(1956), 261-264.

D. W. Boyd and J. S. W. Wong

1. On nonlinear contractions, Proc. Am. Math. Soc. 20(1969), 458-464.

F. E. Browder

1. On the Fredholm alternative for nonlinear operators, Bull. Amer. Math. Soc. 76(1970), 993-998.

2. Normal solvability and the Fredholm alternative for mappings into infinite dimensional manifolds, J. Function. Anal. 8(1971), 250-274.

P. J. Cohen

1. Factorization in group algebras, Duke Math. J. 26(1959), 199-205.

L. Collatz

1. Functional analysis and numerical mathematics, Acad. Press, 1966.

H. S. Collins and W. H. Summers

1. Some applications of Hewitt's factorization theorem, Proc. Amer. Math. Soc. 21(1969), 727-733.

P. C. Curtis, Jr. and A. Figà-Talamanca

1. Factorization theorems for Banach algebras, in Fucntional algebras, edited by F. J. Birtel. Scott, Foreman and Co., Chicago, Ill., 1966, 169-185.

J. Daneš

1. A geometrical theorem useful in nonlinear functional analysis, Bollet. Un. Mat. Ital. (4) 6 (1972), 369-375.

R. S. Doran and J. Wichmann

1. Approximate identities and factorization in Banach modules, Lecture notes in Mathematics, Springer-Verlag, to appear.

H. H. Ehrmann

1. On implicit function theorems and the existence of solutions of nonlinear equations, MRC, Univ. of Wisconsin, Madison, 1962, Report 343.

M. K. Gavurin

1. Concerning existence theorems for nonlinear functional equations, Metody Vyčislenii, Vypusk II, 24-28 (Russian) Izd., Leningr. Univ. (1963).

E. Hewitt

1. The ranges of certain convolution operators, Math. Scand. 15 (1964), 147-155.

2. E. Hewitt and K. A. Ross, Abstract harmonic analysis, II, New York·Heidelberg·Berlin, 1970.

E. Hille and R. S. Phillips

1. Functional analysis and semigroups, AMS, 1957.

L. W. Johnson and D. R. Scholz

1. On Steffensen's method, SIAM J. Num. Anal. Vol. 5 (1968), No. 2, 296-302.

L. V. Kantorovich and G. P. Akilov

1. Functional analysis in normed spaces, Pergammon Press, New York, 1964.

T. Kato

1. Nonlinear evolution equations in Banach spaces, Proc. Symp. Appl. Math., vol. 17, 50-67, Amer. Math. Soc., Providence, 1965.

W. A. Kirk and J. Caristi

1. Mapping theorems in metric and Banach spaces, Bull. Acad. Polon. Sci. (to appear).

L. Kivistik

1. On a class of iteration processes in Hilbert space (Russian), Tartu Riikl. Ve. Toimetised 129(1962), 365-381.

Konrad Knopp

1. Theory and applications of infinite series, Blackie and Son, London and Glasgow, 1944.

J. Kolomý

1. Normal solvability, solvability and fixed-point theorems, Colloquium Math. 29(1974), 253-266.

A. I. Košelev

1. On the boundedness in L_p of the derivatives of the solutions of elliptic differential equations, (Russian) Matem. Sbornik, 38 (80), (1956), No. 3, 359-372.

M. A. Krasnosel'skii and S. G. Krein

1. Iteration process with minimum residua (Russian), Matem. Sbornik (NS), 31(1952), 315-334.

M. A. Krasnosel'skii, G. M. Vainikko, P. P. Zabreiko, Ya. B. Rutitzkii and B. Ya. Stetzenko

2. Approximate solution of operator equations, (Russian) "Nauka", Moscow 1969.

N. S. Kurpel'

1. Projection-iterative methods of solving operator equations (in Russian), "Naukova Dumka", Kiev, 1968.

P. Laasonen

1. Ein überquadratisch konvergenter iterativer Algorithmus, Annales Acad. Sci. Fennic., Ser. A, I. Mathem. 450(1969), 3-10.

J. Moser

1. A rapidly convergent iteration method and nonlinear partial differential equations - I, Ann. Scuola Norm. Sup. Pisa 20 (1966), 265-315.

I. Mysovskih

1. On convergence of Newton's method (Russian), Trudy Mat. Inst. Steklov 28(1949), 145-147.

J. Nash

1. The imbedding problem for Riemannian manifolds, Ann. Math. 63 (1956), 20-63.

M. Z. Nashed

1. Differentiability and related properties of nonlinear operators: Some aspects of the role of differentials in nonlinear functional analysis and applications, edit. by Louis B. Rall, Academic Press, New York·London, 1971.

J. M. Ortega

1. The Newton-Kantorovich theorem, Amer. Math. Monthly 75(1968), 658-660.

J. M. Ortega and W. C. Rheinboldt

2. Iterative Solutions of Nonlinear Equations in Several Variables, Acad. Press, 1970.

W. V. Petryshyn

1. On the approximation-solvability of equations involving A-proper and pseudo-A-proper mappings, Amer. Math. Soc. 81(1975), 223-312.

S. I. Pohožaev

1. Normal solvability of nonlinear mappings in uniformly convex Banach spaces, Funktsion. Analiz. i Ego Priložen., 3(1969), No. 2, 80-84 (Russian, see English translation).

2. On nonlinear operators having weakly closed range, and quasi-linear elliptic equations, Mat. Sbornik 78(120)(1969), No. 2. English transl. Math. USSR Sbornik, vol. 7(1969), No. 2, 227-250.

P. H. Rabinowitz

1. Periodic solutions of nonlinear hyperbolic partial differential equations II, Comm. Pure Appl. Math., 22(1969), 15-39.

H. Reiter

1. L^1-Algebras and Segal algebras, Lecture Notes in Mathematics, Springer-Verlag, 1971.

W. C. Rheinboldt

1. A unified convergence theory for a class of iterative processes, SIAM J. Numer. Anal. 5(1968), 42-63.

J. Schauder

1. Über lineare elliptische Differentialgleichungen zweiter Ordnung, Math. Zeitschr. 38(1934), No. 2, 257-282.

J. W. Schmidt

1. Konvergenzschwindigkeit der Regula Falsi und des Steffensen-Verfahren im Banachraum, Z. Ang. Math. Mech. 46(1966), 146-148.

J. Schröder

1. Nichtlineare Majoranten beim Verfahren der schrittweisen Näherung, Arch. Math. 7(1956), 471-484.

J. Schwartz

1. On Nash's implicit functional theorem, Comm. Pure Appl. Math. 13(1960), 509-530.

Lloyd L. Smail

1. Elements of the theory of infinite processes, McGraw-Hill, 1923.

S. L. Sobolev

1. Some applications of functional analysis in mathematical physics, Izdat. Leningrad. Go. Univ., Leningrad, 1950.

T. B. Solomyak

1. Solution of the first boundary problem for quasi-linear equations of elliptic type containing power-type non-linearities. (Russian), Dokl. Akad. Nauk SSSR, 127(1959), 274-277.

S. Yamamuro

1. Differential calculus in topological linear spaces, Lecture Notes in Mathematics, Springer-Verlag, 1974.

T. Yoshimura

1. Descending problem in Green's function approach to Quantum field theory, Commun. Math. Phys. 40(1975), 259-272.

P. P. Zabreiko and M. A. Krasnosel skii

1. Solvability of nonlinear operator equations, Funktsion Analiz i Ego Priložen 5(1971), 42-44.

D. Zinčenko

1. Some approximate methods of solving equations with nondifferentiable operators (Ukrainaian), Dopovidi Akad. Nauk Ukrain. RSR(1963), 156-161.

INDEX